Connolly

systems analysis

Designed by
Judith Olson

systems

analysis

Edited by
Alan Daniels
formerly head of the Education Department,
National Computing Centre, Great Britain
and
Donald Yeates
head of Professional Education in Systems and Programming,
National Computing Centre, Great Britain

American Edition published by

Science Research Associates, Inc., College Division
165 University Avenue, Palo Alto, California 94301

A Subsidiary of IBM

American edition published 1971 by Science Research Associates, Inc.

SBN: 273 42 064X
Original title: *Basic Training in Systems Analysis*
© National Computing Centre 1969
Adapted by permission of Sir Isaac Pitman and Sons Limited.

preface

This edition has been adapted to conform to current American business practices and data-processing terminology. In addition, sets of problems and exercises appropriate for student work have been added at the end of each chapter.

Preface to the British Edition

This book is based on the National Computing Centre's Systems Analysis Package, and it may be helpful to give some account of how the Package was evolved and the kind of thinking that inspired it.

In October, 1965, an interdepartmental working group was set up by the Department of Education and Science in consultation with the Ministry of Technology and other Government Departments concerned, to consider what steps could be usefully taken in the education system to improve the supply of trained personnel working with computers, in the light of expected demand for such personnel. Following the publication of the group's report in 1967 it was apparent that there was going to be a considerable shortage of systems analysts by 1970.

It was against this background that the National Computing Centre established, under the direction of Professor G. Black and Mr. R. McQuaker, a working party consisting of computer manufacturers, computer users, and consultants to investigate the situation and offer a realistic solution to the immediate problem. It seemed that there were, in fact, two problems: first, the short-term production of courses to meet the deficiency, and second, the long-term establishment of courses within the schools and colleges to ensure the continual production of computer personnel with the correct balance of data-processing experience.

The working party gave a great deal of attention to the short-term problem in an attempt to produce immediate action, while bearing in mind the requirement for a longer and more extensive training scheme integrated with the national educational system.

It was realized that the systems analyst's job required a great deal of business and computer experience and that this might involve years of training for the complete novice. The short-term approach was therefore based on the recruitment of trainee analysts from work most closely allied to the computer systems field, i.e., computer programmers or men from a commercial environment who had been involved in work requiring fact-finding and analysis in creating new business systems (e.g., organization and methods personnel). This presented the problem that the former have very little commercial experience and the latter have no computer experience. The working party was agreed that a large part of the training must be acquired in the environment of the installation and that formal instruction should be given in periods of time acceptable to the employer. This ruled against the large block course and favoured the modular concept which the working party adopted.

The total training recommended was split into—

1. Two parallel preliminary modules of one to two weeks each;

2. A main module of six weeks; and
3. Advanced modules of three days to two weeks depending on the subject.

It was expected that the preliminary modules would bring both areas of recruitment to a level of assumed experience that would allow them to benefit from the training given in the main module. At the end of the main module the trainee should then be able to take his place in a systems team and work, under supervision, on tasks that would consolidate the initial training. After a period of approximately twelve to eighteen months in the environment of the installation the trainee could return for the advanced modules.

It appeared that although the need had been properly assessed for the development of potential systems analysts, there was another short-term problem which had to be overcome.

Large numbers of systems analysts would be trained through a combination of sandwich courses with the technical colleges and "in company" training through company training schemes. The main module of the course of training outlined by the report required an experienced systems analyst to interpret the headings provided in the syllabus so that they would be implemented to maximum effect.

The technical colleges did not contain large numbers of teachers with practical qualifications in systems analysis and the practising systems analysts within company training schemes did not have the time to analyse their skills and present them through suitable training material. The working party therefore suggested the development of a "package" of training material built up of suitable case studies, lecture notes, and visual aids around which the technical college teacher or the practising systems analyst could build a course to interpret the syllabus of the six-week module. This material was developed on the basis of job-oriented training.

Most training schemes develop through many educational phases based on specific aspects of the main functions involved in the job and confuse the learner who cannot see the final job function. Therefore the first aspect of the material was designed to illustrate clearly the functions of the systems analyst so that the learner achieved a feeling of job understanding and job mastery before delving deeper into the information which gave him the overall picture of the system he had been examining. This section was consolidated by a case study which provided most of the facts and formats required to design a system.

The second phase of development probed more deeply into fact-finding, fact analysis, form design, etc., and was consolidated by a second case study.

The third phase of development took in all the aspects that had not developed to date and added the physical and human problems of implementation. The course was completed with a case study which contained all aspects of these three phases.

At one stage during the development it was suggested to the test group using the material that the course should be based on one complete case study. This was rejected by the group. They believed that the three complete case studies summarized the main procedures at each stage and gave greater insight into the total concept of Systems Analysis.

In this way the learner moves from a confident feeling of understanding and job mastery to a development of individual skills in the job function. At each stage he is motivated by the phase ahead of him and sees the relationship of the new phases of learning to the total job function.

To a certain extent this is a development of Thomas Gilbert's work on skills training but the Package itself has provided an example of how effectively main core material can be developed, when account is taken of educational theory and practice, and all relevant media are integrated fully, so as to achieve a total learning system.

Alan Daniels
Donald Yeates

acknowledgements

The N.C.C. Systems Analysis Package was the work of many hands. The Editors are pleased to have this opportunity of thanking individuals and organizations who contributed.

Thanks are owing firstly to the members of the N.C.C.'s working party. They were—

Professor G. Black (Director) *Chairman*	The National Computing Centre
Mr. R. J. McQuaker *Vice-Chairman*	The National Computing Centre
Mr. E. J. Howe *Secretary*	The National Computing Centre
Mr. V. S. Baker	Central Electricity Generating Board
Mr. E. Bird	I.C.T. Ltd.
Mr. S. E. Bird	Department of Education and Science
Mr. J. A. Brunt	English Electric Computers Ltd.

Mr. W. S. Bainbridge	Formerly I.B.M. United Kingdom Ltd.
Mr. M. Crowther-Watson	I.C.T. Ltd.
Mr. T. Crippin	The Scottish Education Department
Mr. W. A. Donaldson	University of Strathclyde
Mr. C. B. B. Grindley	Urwick Diebold Ltd.
Mr. J. E. Hill	N.C.R. Ltd.
Mr. P. Jeffreys	Management Dynamics
Mr. Lyndon A. Jones	S. W. London College
Mr. E. C. Lay	British Computer Society
Mr. P. J. Maylor	N.C.R. Ltd.
Mr. R. J. Philo	Post Office Head Quarters
Mr. F. L. Pitt	Rockware Glass Ltd.
Mr. B. G. Roberts	Belling & Lee Ltd.
Mr. R. K. Stamper	Ashorne Hill College
Dr. S. Summersbee	Brighton College of Technology
Mr. C. A. Stebbing	London Borough Management Services
Mr. R. P. Tomlin	I.C.T. Ltd.
Mr. L. J. Trott	English Electric Computers Ltd.
Mr. D. G. Toose	Department of Education and Science
Mr. P. H. Vince	I.B.M. United Kingdom Ltd.

The following served on a Committee nominated by the Department of Education and Science to give the views of Colleges of Further Education—

Mr. D. Crump	City of Birmingham College
Mr. J. C. Davies	City of Westminster College
Mr. A. H. Jones	Mid-Essex Technical College
Mr. R. K. McEntegart	Manchester College of Commerce
Mr. R. Penny	S. W. London College
Dr. S. Summersbee	Brighton College of Technology
Mr. S. Waters	Woolwich Polytechnic

To obtain guidance on the suitability of the Main Course Syllabus before the preparation of teaching material, a questionnaire was sent to a number of organizations. The following are thanked for their response, and for their additional helpful comments and suggestions—

A.E.C. Limited
Babcock & Wilcox Ltd.
Beecham Group Ltd.
Castrol Ltd.
C.A.V. Ltd.
Central Electricity
  Generating Board
Esso Petroleum Ltd.
Hampshire County Council
The Institute of Chartered
  Accountants of Scotland
International Harvester Co.
International Publishing
  Corp.
S. C. Johnson & Son Ltd.
The Klinger Manufacturing
  Co.
Legal & General Assurance
  Society

Leicester Regional College of
  Technology
The London Hospital
Parkinson Cowan Appliances
Peat Marwick Mitchell & Co.
Godfrey Phillips Ltd.
Post Office
Pressed Steel Fisher Ltd.
Rowntree & Co. Ltd.
Royal Insurance
Salisburys Handbags Ltd.
South Western Gas Board
Tesco Stores Ltd.
The Thomson Organization
Touche, Ross, Bailey, Smart
  & Co.

The following people were involved, on a full or part time basis, in developing a large part of the course material and presenting the Pilot Course—

Mr. B. Sawyer
  (Course Director)
Mr. G. Wood
  (Course Director)
Mr. D. Favre
Mr. K. Fraser
Mr. P. Jeffreys
Mr. G. Ling

Mr. B. Sawyer

I.C.T. Ltd.

S. W. London College

N.C.R. Ltd.
Management Dynamics
Management Dynamics
English Electric Computers
  Ltd.
I.C.T. Ltd.

Mr. M. Smith	Ashorne Hill College
Mr. R. Stamper	Ashorne Hill College
Mr. R. Waller	I.B.M. United Kingdom Ltd.

Finally a special tribute is due to Mr. Ron Negus, who prepared the manuscript of the present volume. Out of the mass of material available for the Package (including, it should be said, the most recent revisions) he has distilled the essence. The final form of the text owes a great deal both to Mr. Negus's technical knowledge and to his editorial skill.

A.D.
D.Y.

contents

illustrations

1

the scope of systems analysis

Before one can study the techniques a systems analyst must understand and apply in his work, one must appreciate his function in the wider context of the organization in which he is employed. Regardless of what purpose the organization serves, a systems analyst must have certain fundamental personal qualities if he is to succeed in his chosen vocation.

This book is primarily directed toward new entrants to the field of systems analysis. The following brief discussion of the analyst's purpose is intended to establish a basis of agreement about what analysts seek to be and do. This definition of purpose will help the new analyst see his total activity in perspective before he becomes immersed in the finer details of his new job, which are described in subsequent chapters.

Personal qualities and training 1.1

The disadvantage of setting down a list of qualities required for any job is that no one person possesses them all. However, it is worthwhile to recognize the attributes that the job demands, since the trainee will probably acquire those he lacks as his experience grows.

An analyst must be able to discover the fundamental logic of a system, to produce sound plans, and to appreciate the effects of new facts in his planning. He must be perceptive, but must not jump to quick conclusions. He must be persistent to overcome difficulties and obstacles, and maintain a planned course of action in spite of setbacks.

The systems analyst needs stamina, strength of character, a sense of purpose, and a broad flexible outlook. An orderly mind, a disciplined approach, and logical neatness are necessary because his job will frequently require working without direct supervision.

He must be able to express his thoughts, ideas, and proposals clearly, both orally and in writing. He must also be a relaxed listener. To maintain control through numerous interviews, he must be an accurate and precise conversationalist and must have more than average social skill to communicate and work with others.

Suitable candidates for systems analysis trainee positions often come from a background of analytical work in the commercial area or computer-oriented work involving programming. Since the systems analyst will need both types of experience, the first step in training must be to develop the area of experience that the trainee lacks. Some trainees must develop communication skills, both oral and written; others must be trained in application areas. The level of experience to be developed may vary from a restricted and detailed knowledge of a specific area such as steel production control to an overall understanding of a broad area such as management information systems.

This area of training is similar to an apprenticeship and cannot be satisfied, as has sometimes been tried in the past, by a two-week course in business organization. The more successful analysts have served lengthy apprenticeships in an application area in addition to their professional qualifications.

The analyst must also be trained in the use of such relevant hardware as computers and peripheral equipment. He must know the types of equipment best suited to the solution parameters, overall methods of computer application strategy, and detailed computer systems techniques. Training in this area must be continuous, because new equipment and techniques constantly arise.

Finally, the analyst must be trained in the appreciation of software. He must know what packages are currently available, since he must decide which are suitable for his needs. He must also understand the principles of programming. It is pointless for him to specify a system that cannot be programmed with currently available software.

Organizational background 1.2

Systems within an organization do not exist in a vacuum. They reflect the organization's structure and purpose, and to a lesser degree are influenced by the personalities of both management and staff concerned. When the systems analyst is to start work on a particular project, he must have an informed awareness of his employer's organization. This will help him during employee interviews and will also assist him in placing details of the current system in perspective. Often insufficient time is devoted to this area of training. As a result, the analyst does not fully comprehend the significance of some matters discovered during the fact-finding phase of his work, and he may design the new system with the limitations of the old.

He must also understand the constitution of the organization. This topic covers the overall arrangement of its constituent parts, and the strength of its affiliations with other organizations. When the organization involved is diversified across more than one location, the analyst should discover the function of each.

The analyst will benefit from a knowledge of the organization's history. By tracing the major milestones in the evolution of the firm, he can see the type of management decisions that were made in the past. He must keep the management's overall policies in mind. He should examine the company's annual reports, employee handbooks, the trade literature, investment reviews, and press clippings. This background will help him appreciate the precise divisional structures currently existing at each location. From this he can discover how much of the structure is carefully planned and how much is dependent on former structures. At the same time he should try to discover how closely the present objectives of each main function are in accord with the policies now being followed.

As part of his fact-finding operations, the analyst will need to detail the departmental structure of the area in which he is working. If there is a company organization chart, the analyst should ascertain that it reflects current conditions and then acquaint himself with it. If no chart exists, he should prepare one. In any case, some clear distinction should be drawn between those operations that involve line management and those that are staff functions. The former are responsible for executing policies and translating them into attainable objectives, while the latter usually act as advisors.

Conventional organization charts, which show a number of branches from a chief executive or board of directors at the top, represent the pyramid of control found in most organizations. The analyst should superimpose on this structure another one that describes the true lines of communication. This structure will not follow the strict divisional and departmental boundaries implied by the original chart, but it will show the analyst how the proposals he may make in one area may affect the work in others. He will probably find much of the material mentioned above recorded in one form or another, and will have to edit and distill it to obtain a coherent picture of the organization. However, he will probably not find any recorded information about the environment in which the organization operates; he will need this information also. The firm itself has little direct influence on its environment, which is constantly changing.

Some examples of facts the systems analyst should try to discover are the organization's position within its industry, the labor relationships in the industry, and the effect of government policies on the industry. A new system may operate no better than a previous system, although it may appear much better on paper, simply because some environmental condition frustrates its operation. The systems analyst may find it necessary to estimate and predict environmental conditions to anticipate flexibility of the system he designs.

1.3   Human aspects of the job

The analyst must constantly remember that the systems he is designing affect people. He must understand possible reactions to his system and the reasons behind them, even if they seem irrational.

We all have our systems of belief that cause our viewpoints to differ. They are based upon our experience, education, emotional makeup, intelligence, knowledge of specific areas, and our particular interests. Highly intelligent, logically-minded people may have radically different yet reasonable views on the same subject. Using the same data, they have reached these views because their basic premises differ. These premises have been developed through the years, and are based on assumptions about the many nonmeasurable factors in our environment.

Very few people are both highly intelligent and coldly logical. Our emotions are close to the surface and have a profound effect on our thinking. We tend to believe what we want to believe. Some of the illogicalities that are normal include generalizing from particular cases, assuming cause and effect from a correlation, attributing a logical reason to an emotional belief, and transferring subconscious dislikes to people or things our conscious minds will accept as scapegoats.

The main motivating forces in mankind have been identified as —

1. Physiological needs such as those for food, shelter, rest, variety, safety, and sex;
2. Safety and security against danger and loss of physiological satisfactions;
3. Social needs such as the needs of association, acceptance by associates, giving and receiving friendship;
4. Ego needs such as recognition, sense of achievement, prestige, independence, realizing one's own potential, and being creative.

Other motivating factors are curiosity, which often causes people to look for change or new ideas, and habit persistence, or preferring the familiar to the new. Habit persistence is associated with problems of unlearning. It is difficult to change automatic reactions to external stimuli.

Most people have achieved a reasonable degree of satisfaction in meeting the above motivating needs. Fear of losing them is a powerful resistant to change. Helping people meet these needs has been a basic approach in motivation research and salesmanship for many years. However, most people resist this approach when they realize it is being used.

The analyst must persuade others to overcome resistance to change. He should gather as much information as possible about the system of beliefs of the person to be persuaded, and about the objections he may raise. Careful observation, including attention to expressions and casual words, may help detect what a man is trying to conceal. If a group rather than an individual is concerned, the analyst should identify the most influential person in the group, and try to persuade him. He should establish himself as a person who can be trusted and persuade others of his sincerity and concern for their interests.

He should plan his approach and determine his intermediate objectives, considering all possible objections and the alternatives that can be offered. Timing should be such that the degree of pressure will be correct. He must plan how to get others to criticize their present situation or the products they are using and create a desire for change. The following suggestions may be useful.

1. Do not expect instant conversion.
2. Induce participation in decision-making; this causes commitment to the decision.
3. Avoid using too many arguments by emphasizing essentials.
4. Ask questions that will emphasize areas of agreement.
5. Aim at a mutually satisfactory solution, not at total conversion.
6. Avoid any criticism of the past; concentrate on positive aspects of the change and a common desire to make progress.
7. Listen sympathetically to problems and objections but do not assume that verbal objections are necessarily the real ones; there may be rationalizations of emotional objections that the person knows are irrational and will therefore not admit.
8. Sell alternatives, such as "If you buy one would you prefer model x or model y?"
9. Be wary of negative suggestions, such as "It would be fine if we could do so and so, but it is not really possible in a firm like ours." This approach may cause an aggres-

sively favorable reaction to the implied challenge, but if it fails, it creates a situation from which it is difficult to recover.

Assume that acceptance has been won and stress the advantages and benefits while admitting the difficulties. Suggest that management and supervision will be able to overcome the inevitable transitional problems and make the new system work.

Ask for suggestions as to how the more obvious problems may be overcome. Emphasize continued support and help both during and after the change. Once agreement seems reasonably certain, get the plans, including a program and timetable, accepted quickly. Give management and supervision the credit for the change. Remember that to be logically right can be psychologically wrong.

Some ground rules for resisting persuasion are worth considering, since they indicate what the persuader must overcome. They are —

1. Don't listen.
2. Attribute ulterior motives to the persuader, preferably behind his back to others who may be affected.
3. Concentrate on disliking him.
4. Exaggerate objections, especially the danger of repercussions and the unsuitability of timing.
5. Raise the temperature of the discussion, and discuss personalities wherever possible.
6. Keep your real objections to yourself.
7. Stick to your prejudices.

The analyst must consider the possible reasons employees resist change. They are —

1. Fear of losing the job, of wage reduction, of inability to learn a new job, of loss of prestige, of loss of interest in the job;
2. Suspicion of management's motives in making the change;
3. Resentment against personal attack, or a feeling that any change is a personal criticism of the way a man was doing a job;
4. Social upset caused by breaking up a working group;
5. Ignorance, or fear of the unknown.

Among the ways of overcoming these background reasons for resistance are—

1. Keep people in the picture well in advance, give the full reasons, and sell the benefits.
2. Give people an opportunity to participate by making suggestions.
3. Give security, which may mean guaranteeing the financial future or providing retraining.
4. Take time in introducing change; create a favorable atmosphere; give people time to get accustomed to an idea before implementing it. (There are, of course, rare occasions where the quick introduction of change without notice may be best.)
5. Provide sound personal examples. Much depends on the degree of confidence employees have in management.
6. Cultivate the habit of change. If changes are frequent, people will be more used to the idea, and changes will be more readily acceptable.

1.4   The design process

For various reasons, it is not practical to design a company's total system as one major short-term project for the computer. The company cannot accept a massive one-time change, and management needs time to define problems. The shortage of skilled systems analysts also prevents the adoption of the total-change approach. Management must therefore divide the total company system into subsystems and provide the systems analyst with their details.

In most commercial organizations that do not have computer-based systems, throughput of data tends to be chronological. The input to any one department occurs in cycles. As each cycle is received, it is processed against manually maintained data files. Additions, selection of information, and summaries are made for management control and updating of the files.

The input, having been processed in the department, is then passed for action some time later in the day. In subsequent departments, similar additions, selections, and summaries are made. This process continues; the next cycle of input occurs.

The information abstracted from data flowing in this way tends to have a chronological bias, and the importance of the selection is not recognized.

In integrated computer-based systems, the input is entered once. The outputs, in the form of selections and summaries, are selected because of their *importance* rather than on a chronological basis. A systems analyst designing an integrated computer application must recognize the importance of information to be selected from the system.

The large majority of computer installations work with input provided at the supervisory level, both in terms of the internal and the external environment. For example, internal environmental input may be factory notifications of finished stocks, interdepartmental transfers, or internal accounting adjustments. External environment input consists of data such as customer orders, invoices, and tax data for payroll.

If the computer is used at this level with these inputs, its outputs can provide the next level of management with information such as historical analyses, summaries, and internal file data. The outputs provided to the external environment are similar to the inputs, such as invoices, statements, and checks.

If a company intends to use the computer to perform tasks that will directly help management form policy, the inputs to the system must enable the computer to provide probability statistics. At this level, the internally provided company inputs will consist of as much of the company's history, including all aspects of the total system, as can be accumulated. External factors about economic trade and local conditions must be added to make a comprehensive management information system.

In practice, the areas in which a computer is employed are never as clear as those described above. It is quite conceivable that a small computer can provide top management with some information, and, conversely, that a large computer can easily handle the day-to-day data-processing work of the company. However, the help a computer can give management is only that of a tool whose main function is to provide feedback from the company's activities. Management can establish much closer control over the future use of company facilities. Future computer applications will provide the company policymakers with much more information, so that a more coherent plan of action may be passed to the chief executive.

After the management requirements of a system are established, the analyst must generate a set of descriptions to explain how outputs are derived from inputs. Management requirements will normally be output oriented. They will have to be refined, and further outputs will need to be determined.

The analyst may be involved in determining the inputs that are required. Once input information is captured by a system, it must be preserved in ordered files of records. The analyst must determine what files are needed and what records should be kept. In order to obtain the required output, he must specify what mechanical and clerical procedures will be required.

The design sequence is therefore—

1. Outputs (results)
2. Inputs (data)
3. Files (files)
4. Procedures (program)

Stages 2 and 3 may have to be resolved simultaneously. Similarly, stages 3 and 4 are interlinked.

In general, this simple sequence must be followed several times. At each stage, decisions that will affect subsequent stages are made. It may be necessary to change decisions made in earlier stages for reasons of technical feasibility, programming complexity, operating cost, and so forth. Thus the process is one of successive repetition of the design procedure until a satisfactory solution to the overall problem is found.

Although a major part of the system is defined by the computer programs that implement it, even well-annotated programs give only an understanding of the programming mechanics. The system specifications are absolutely essential; they alone constitute the definition of the system. As such, they are the major means of communicating the project's requirements to the programming and operational staff.

It was stated above that the design process is iterative; that is, that successive refinement of the design is required to generate the final acceptable solution. This implies that at each stage of development the design must be examined to check its adequacy. It will ultimately be judged by the user-management and by the chief systems designer; they will also want to assess the adequacy of the design.

The criteria of good system design may be summarized as follows:

1. Management objectives realized;
2. Well-defined computer system;
3. Well-designed human aspects;
4. Efficient and timely operation;
5. Carefully planned and tested implementation;
6. Accurately estimated costs; and
7. Rigorous design methodology.

The various aspects of this overall process are discussed in the following chapters. A careful application of the principles shown will help the analyst produce good systems designs consistently.

Exercises

1. What background experience do systems analysis trainees frequently have?

2. Which of your own personal qualities do you consider will be assets in systems analysis work?

3. Why is it important for an analyst to understand an organization's history, structure, and goals? Discuss this in terms of an organization with which you are familiar.

4. Why do some people in an organization tend to resist change? What techniques might a systems analyst use to overcome this resistance? Discuss a business situation in which you have had to overcome such resistance.

5. Assume you are a systems analyst assigned the task of designing a new system for accounts payable, which is presently done manually. The person responsible for it for the past twenty years is a conscientious woman in her early fifties who has earned a reputation for competence and is very proud of the fact that her records are always current. As the volume of transactions has

grown over the years, she has worked overtime more and more often rather than hire an additional person who might, as she puts it, "gum up the works." What would you expect her attitude toward a new system to be? What points would you plan to discuss with her during your initial interview to gather facts about the application?

6. In designing a computer-based system for an organization, what design sequence will the systems analyst follow? What criteria must the final design meet?

2

systems investigation

Any investigation of an existing system in an organization begins by establishing a plan to show where the investigation is to be made, what methods are to be used, and what schedule is to be followed. Analysts will then be assigned to the investigation. One may be given the entire responsibility for a small assignment, but on larger jobs, the task will probably be divided among several analysts. The overall plan should indicate points where special members of the analysis team may be required.

After the planning and staffing is completed, each department to be investigated must be notified. They should be told the general terms of the investigation, which analysts will be conducting it, and how long they are expected to be in each department. Each analyst can then proceed with his investigation, which will consist of two stages: fact-finding and analysis. Fact-finding involves establishing precisely what systems are currently operating in the area under investigation. Analysis is concerned with recording the facts to display the logical structure of the system, so that its purpose can be clearly understood.

When the investigation is completed, the analysis can be studied in relation to the system's management objectives. The analyst will usually begin a design of the new system at this stage. The design and implementation aspects of the job are the subject of subsequent chapters. This chapter will introduce some basic concepts about fact-finding and recording.

2.1   Management audits

Sometimes an analyst is required to perform a management audit. In this case, the fact-finding and analysis stages are completed as before, but the prime goal of the work is to report on the efficiency of an existing system rather than to design a new one. For example, after the introduction of some new office machinery, a management audit might be carried out.

The analyst should develop a checklist to discover how the organization works. Such a list covers most areas he must analyze. However, if it is used carelessly, there is a danger that only the questions appearing on the list will be asked. Structural problems may be overlooked. In all cases where checklists are developed, two final questions should be added: What else must I look for? What have I missed?

Following is an example of a checklist used to examine a purchasing department.

PLANS AND OBJECTIVES

1. Have definite plans and objectives for the department been established?
2. Are the plans and objectives of the purchasing department in harmony with those of other departments as well as with the company as a whole?
3. Has adequate time been allotted for planning and determining better ways of meeting objectives?
4. Is there a clear understanding of the soundness and practicality of objectives?
5. Does top management agree with the purchasing department's plans and objectives?

6. What points should be considered to improve the purchasing department's plans and objectives?

ORGANIZATION STRUCTURE

1. Is an organization chart available and maintained currently? (If not, the analyst should prepare one.)
2. Is the organization structure sound and effective?
3. Does the organization reflect the program and objectives?
4. Are the various duties and responsibilities delegated properly and defined clearly?
5. Are the lines of authority effective?
6. Is there overlapping or duplication of functions?
7. Can any organizational elements or functions be eliminated or transferred to other departments?
8. Can the organizational setup be changed to increase coordination of activities?
9. Are the functions assigned to key people in proper balance?
10. Are the various functions coordinated?
11. Do the people involved understand their responsibilities and authorities?
12. What steps should be taken to make the organization structure more effective?
13. Does the typical employee in the department understand the organization structure?
14. Is there provision within the department for regular reviews of the organization structure?

POLICIES, SYSTEMS AND PROCEDURE

1. How are the purchasing policies determined?
2. Have all purchasing policies been written?
3. Do the purchasing policies reflect the basic objectives and goals of management?
4. Are the purchasing policies positive, clear, and understandable?
5. Are they made known to the department's staff?
6. How are established policies enforced?
7. How are suppliers selected?
8. Is the function of procurement entirely centralized?

9. Do purchasing requisitions show the necessary approvals by authorized persons? Are they within financial limits?
10. How are excess deliveries handled?
11. How are defective items handled?
12. Are all purchasing policies complied with?
13. Is the purchasing system meeting all current requirements and operating effectively?
14. Can the general routine of processing paperwork be improved?
15. Can the system be improved to reduce costs?
16. Are the purchasing procedures written?
17. Have adequate controls of material rejection been established?
18. Has sufficient consideration been given to internal control?
19. What is the general condition of the records?
20. Have definite procedures been established for all functions?
21. Are they fully complied with?
22. Are government regulations met?
23. Has the legality of purchase order terms and conditions been checked?
24. Can any records be eliminated?
25. What specific procedures require immediate study and revision?

PERSONNEL

1. What is the general layout of office space and equipment?
2. Is the office laid out to provide maximum use of space and efficient work areas?
3. Is there an area for reception and interviews with salesmen?
4. What is the general condition of office equipment?
5. Describe all mechanical equipment in use.
6. Is maximum use of the present office equipment made?
7. Is the equipment located where it can be used conveniently?
8. Is adequate storage space provided for?
9. Are the files reviewed regularly for transfer to storage and records retention?

OPERATION AND METHODS OF CONTROL

1. Are management reports adequate, clear, and prompt?
2. Is the normal lead time for procurement generally followed?
3. What are the expediting methods?
4. What safeguards against possible irregularities are established?
5. What has been done to achieve greater standardization?
6. What are the causes of overtime? What can be done to eliminate them?
7. What are the principal means of control?
8. How can the various operations be improved?
9. Are purchase orders placed as the result of competitive bidding?
10. To what degree are the required specifications adhered to?
11. Can any operations be eliminated, simplified, combined, or improved by sequence changes?
12. Are there any bottlenecks? What is being done to eliminate them?
13. Can or should any operations be mechanized?
14. What methods of measuring productivity are established?
15. Are work units identified and standards developed? Are the standards obtainable?
16. Is work-study training needed?
17. Are forecasts of future trends established?
18. Is there budgetary control over all expenditure?
19. Do reports give comparisons with past periods and predetermined objectives?
20. Is there a means of determining the cost variance on material purchases?
21. Has a clerical work-measurement program been established? Is it working?
22. What clerical cost controls should be established or expanded?
23. What is needed to increase purchasing efficiency?
24. What can be done to increase the quality of work?

Fact-finding methods: interviews   2.2

Interviews, questionnaires, and observation and record inspection are techniques that can be used for fact-finding. Interviews are often the most productive fact-finding source, and the analyst should learn how a good interview is conducted. Obviously, he must have a certain amount of skill to meet his objectives. He should be impartial, tactful, and skillful in influencing others to accept advice they would prefer to ignore. Interviewing is the process of obtaining information by means of conversation without upsetting the other party: this entails being a good listener, keeping the conversation rolling, and keeping the subject on appropriate lines. Conversation is itself an art that relies on the ability to suit the treatment of any subject to the person, the place, the mood, and the moment.

Interviewing in systems work has two main objectives: to enable the interviewer to discover and verify facts, and to provide an opportunity to meet and overcome resistance. The first is an obvious necessity: facts must be obtained, and interviewing is sometimes the best way to obtain them. Further, much of an assignment's success can depend on the reputation for being helpful and practical that the analyst creates. Even greater success depends on his ability to work around or avoid resistance from employees whose work habits may be changed.

The analyst will be dealing with a very wide variety of personalities at different levels of authority. He may interview the head of a small department; in a large department, the level at which he opens negotiations may be lower. He will need to adapt his approach, timing, and phrases to suit each person. Often the person in charge will accompany the interviewer on his rounds of the staff. Sometimes the supervisor will attempt to answer all the questions himself, or to conduct the interview. In this situation, the analyst must be tactful, but he must politely and firmly retain control of the interview.

Resistance to change will often produce a somewhat frigid climate for the interview. The more effectively an employee is performing his job, the less likely he is to welcome someone interfering with his methods. Those who are genuinely interested in their results often become wedded to proven methods. They are suspicious of change and may have never considered, or heard of, various alternative machines or processes. Until a reputation for helpfulness and soundness of advice has been earned, the interviewer is often met by suspicion, by reluctance to change, and sometimes by resistance to any suggestion of interference with "my job."

The analyst should remember that systems investigations are concerned with *existing* procedures and well-established habits. He will be working at another person's desk, often in an open office with no privacy, in full view and within the hearing of a number of people. They may be listening to what is happening and may themselves be interested in and affected by the conclusions reached at the interview. This lack of privacy makes the interviewer's approach more important than usual. However, the analyst should remember that slips can be retrieved, and admission of errors that result from misinterpretation of facts or from personal misunderstanding is in some cases preferable to incomplete concealment.

The success of the interview itself is conditioned by at least two variable and probably incalculable factors — the personal prejudices of the interviewer and the person interviewed. Interviewing is not the only means of fact-finding; it should be supplemented by observation and examination. However, it is usually the only means of finding out something about the unseen part of the job, that which goes on inside another's head. And the higher the interviewer goes into an organization, the more important is this side of the work. Therefore, to achieve the best possible results, he must remind himself to eliminate his own bias and be impartial and thorough. He must think about the person he is interviewing and put himself in the other's place. Many people who are very enthusiastic and whole-hearted about their jobs may regard a systems analyst's visit as a suggestion that their work is not well done. The interviewer should recall his feelings when *his* work has been inspected critically and make his approach sympathetic.

Because the analyst is on another's ground and faces considerable actual or potential resistance, tact is essential. It will supplement the more definite process of studying what is actually done, what questions should be asked to complete the story discovered by watching the job, and what meaning to give to the observations made or the answers received. To be tactful, the analyst must make the interview impersonal, keep it objective, and prepare for it beforehand. He should emphasize to the employee that the interview will not affect his present job classification or salary. He should also confine his questions to the actual operations being carried out and avoid comparing one person with another, or asking anything that may reflect adversely on the person being interviewed. If possible, the analyst should encourage the employee to go through the motions of his job in detail so that he can get a picture of what is involved before listening to the employee's description of his job, which often contains a certain amount of subjective opinion.

If the analyst has had wide experience, he is sometimes tempted to jump to conclusions based on problems he has seen in the past. In fact, very few problems are exactly the same. Hence, strict self-discipline or freedom from bias is valuable. He must refuse to guess, to rely too heavily on his past experience, or to prejudge the issue.

He must be fully prepared. He should know the names of the persons in charge and of those with whom he is going to talk, and what part of the company's work each particular job represents. If possible, he should know in advance why each person was selected to be interviewed and identify points where he may need more information about their duties. Sometimes he may be able to formulate questions in advance, but he should phrase them naturally and spontaneously during the interview. Later, when he is actually observing a job, it may be possible to arrange his questions in a logical order so that each follows naturally from the last. This approach helps to keep the interview impersonal.

A subsidiary part of preparation is making arrangements for the visit: for example, finding out whether a general announcement has been made about it, and if so, in what terms; or whether suspicion has been created by what has been said or left unsaid. If no announcement has been made, he must decide such factors as whether he would like to announce himself or whether he would prefer the person in charge to perform the introduction. (Some people consider an introduction by management essential and some think that the initial interview should be informal, on the employee's home ground.) After these preparations and the initial introductions, he should state the reason for his visit, and, if possible, he should also describe the method he will be following.

In actually conducting the interview, he should use simple terms. He and the person being interviewed may use the same words with quite different meanings. His words must convey his exact meaning; if necessary, he must repeat the same idea in different words for clarity.

While keeping the discussion strictly impersonal, the interviewer must try to enlist the employee's interest and cooperation. He should indicate that he is trying to eliminate unnecessary effort, to make the organization work more smoothly, to save time, and to improve service. Most people are interested in at least one of these objectives.

The analyst should express appreciation during an interview whenever possible. (However, he should remember that it is dangerous to approve what he may later have to change.) He should avoid mentioning anything that will diminish the employee's sense of importance, which may already have been threatened by the inquiry. He can reassure him by confessing that he has no first-hand knowledge of the job itself and is therefore at a disadvantage until he has been shown what is done.

Details of manner and politeness should be observed. The analyst should ask permission to smoke, arrange for both parties to be seated and comfortable, and cause a minimum of disturbance to the employee's work schedule. He should remember that affectations are unpleasant, and nervousness is catching. He may find it best to start fairly formally, especially if either of the parties is nervous; later he may get more information by being informal. He must remember that the employee may need help in explaining his work, and should guide him without asking leading questions.

It is not sufficient merely to say "thank you" and leave when all the required information has been gathered. The analyst may have to return at a later date. He should, therefore, establish a friendly and helpful relationship in each interview. The interview should not run too long and waste time; on the other hand, it should not be so brief that it may make an employee feel he has been slighted.

To end the interview, he should recapitulate very briefly the object of the interview and read any notes that he had made. This review should refresh the other person's memory of his earlier statements and give him a chance to correct himself. The analyst may also ask if anything has been missed, thereby encouraging the employee to bring forth a valuable suggestion not previously mentioned. At this final stage, the employee may want to divulge "unofficial" opinions or facts. He or she may have been waiting to say what the organization looks like from underneath, how much time is wasted, and how illogical some procedures appear to those who carry them out. Even if this commentary rambles, the analyst should listen carefully for ideas that may relate to his assignment. He must not, however, unduly encourage such confidences or get involved in personalities by agreeing or commenting. Experience will teach him to weave a little authority into a free cooperative discussion so that the interview ends on the right note.

However, there are no strict rules about interviewing. The possession of tact is a great asset, yet there may be occasions when the interviewer has to stand his ground and point out that his duty is to get certain information and that the other man's duty is to supply it. Only when tact has failed should a "showdown" be considered as a last resort. Reactions to pressure may be unexpected and unpleasant.

During an interview one must distinguish between facts and opinions. Both are necessary, but they will be treated differently at the analysis stage. It is also important to make sure that both sides of a story are heard. (For this reason, it is good practice to interview both the recipient and the originator of a document whenever possible.)

The final important task in an interview is to make sure that all questions are asked at the correct level. Asking a question that an employee cannot answer wastes time. For example, top management is competent to answer questions about policies and general philosophy, and assessment of project support; middle management can answer questions of departmental practices and procedures, bottlenecks, time factors, labor restrictions, and so forth; and at the operator/supervisor level questions of functional problems, communication, job recording, and checking may be asked.

2.3   Fact-finding methods: questionnaires

For certain systems investigations the questionnaire method of obtaining information may be useful. It is most valuable when a very small amount of information is required from a large number of persons or when a systematic study of one unit's entire activities is made. The questionnaire may be the only practical method of obtaining information in a limited amount of time when the investigation covers a large number of facilities in a large decentralized organization. Using a questionnaire may also save valuable interview time by allowing the respondent to assemble the required information before the interview.

However, for most surveys, the use of questionnaires is usually not recommended. People often object to answering numerous, time-consuming, tedious questionnaires and it is difficult to design questionnaires that obtain exactly the information desired. Many of the answers received from elaborate questionnaires are often inadequate, and the interview technique must be used to clear doubtful points. Also, the method is relatively slow because many persons delay answering and returning the questionnaire.

If a systems questionnaire is to be used, it should conform to the principles of good form design. Questions should be clearly worded and free from bias. A covering letter should explain the purposes of the questionnaire and state a reasonable date when the questionnaire should be returned.

Sometimes a survey of a report's distribution and usage is needed. Such a survey enables an organization to include only useful and pertinent information, to eliminate unnecessary recipients, to save time, and to reduce the costs of preparing and distributing the reports. In this case, a questionnaire reading as follows might be sent to each recipient of a report—

We understand that you are on the distribution list for _____ report. We are surveying the distribution of various reports to make sure they contain all the required information, to eliminate unnecessary information, and to ascertain that everyone who needs the information receives a copy of the report. We would like your help in this survey. Please answer the following questions:

1. Do you still require this report?
2. If you do, what information do you use?
3. How do you use it?
4. What additional information should be in the report?
5. How would you use it?
Please return your answers by _____.

If a detailed account of the activities of all individuals in a section is needed, the analyst may use a duties questionnaire in which each employee is to list his duties and the average amount of time required for each. An example is shown in

Figure 2-A. Also, it may be helpful to ask for a list of forms and reports connected with the employee's work, machines used in performing his duty, and unusual duties that seldom occur and therefore cannot be conveniently listed among regular duties. The analyst should compare these duty lists with existing job description specifications.

2.4   Fact-finding methods: observation

Being able to observe an operation and draw useful conclusions from the task seems to be an inherent ability that is very difficult to teach or learn. Often those who can observe acutely seem to have a sixth sense to those who cannot. However, observation depends to a great extent on powers of concentration.

FRIENDS INSTRUMENTS, INC.
DUTIES LIST

Last name and initials		Date form completed	
Job title		Department	Section

Enter each main duty you perform, and indicate how many hours per week it requires.

No.	Description of duty	Approx. hours per week

Other activities (lunch, coffee-breaks, etc.)
TOTAL HOURS WORKED PER WEEK

WHEN COMPLETED HAND THIS FORM TO YOUR SUPERVISOR

Fig. 2-A. Duties questionnaire

An analyst usually makes subconscious observations of the work environment as he visits various locations. These small and unexpected observations may later become significant. For example, an analyst might be asked to investigate delays in production caused by subassemblies arriving late from stockrooms. During his investigation of the mechanics of the ordering system, he discovers that a release order prepared on a spirit duplicating machine is required to withdraw the goods. He observes that the bottleneck is in the duplicating section. On further probing, the section supervisor tells him that this bottleneck is caused by pressure of other work. During this interview, he observes that the two duplicator operators seem to be spending most of their time on the production of stencil-duplicated work. Subconsciously he notices that both operators are wearing attractive dresses. After analyzing the situation, he decides that there is sufficient time and equipment in the duplicating section to prepare the necessary documents on schedule. He asks himself why there should be delays. A second visit to the section to pursue a theory based upon his observations elicits the real reason. Both operators consider the spirit duplicators dirty. Not wishing to spoil their clothes, they put off any jobs on the machines as long as possible. Two stylish nylon coveralls may solve the problem and clear the production delays.

The analyst may also carry out intensive observations. This technique, which is particularly good for tracing bottlenecks and checking facts, involves watching an operation to see for oneself exactly what happens. However, long periods of time should not be devoted to this method since it is unnerving to those observed.

A related technique, in which an operation is observed at predetermined times, is called systematic activity sampling. Observation times are chosen at random, and persons do not know in advance when they will be under observation. This technique should not be confused with statistical sampling, although some ideas are common to both. The analyst may use statistical sampling methods when he is conducting record inspection.

Record inspection involves inspecting the actual results and records of the system under investigation. It requires detailed counting of entries, numbers of documents, timing of throughput, and so on, usually to qualify facts obtained at interviews. In record inspection, the analyst should notice trends. For example, he should ask whether the workload is increasing or decreasing, and if so, by how much per month or year. He should also discover ratios such as the proportion of credit notes to sales invoices produced during a normal working day.

Time will often prevent the analyst from investigating records as thoroughly as he might wish. Consequently, he must attempt to draw conclusions from a sample. This approach can be satisfactory if the analyst understands the concepts of statistical sampling. It is a very hazardous exercise for the untrained. One common fallacy is to draw conclusions from a nonrepresentative sample. He should be very careful in estimating volumes and data field sizes from sample evidence. For instance, assume that cash receipts are sampled. A small sample taken during a mid-month slack period might indicate that about 40 receipts are processed per day and that $3600 is the maximum value for any one item. Indiscriminate use of such a sample for systems design might be disastrous. For example, the real situation may be that receipts range from 20 to 2000 per day depending on the time of the month, and that sometimes checks for over $250,000 are received.

Sampling is too large a subject to be dealt with in depth in this book. The analyst contemplating using samples in factfinding is strongly advised to consult a statistician or to study the subject thoroughly.

2.5   Introduction to standards

The nature of the analyst's job requires him to collect large numbers of facts that must be documented in an orderly way. An analyst working in an efficient computer installation will be provided with a set of data-processing standards to follow in describing his system to all levels. He is a vital intermediary in the communications structure. He must report *upwards* to the chief analyst and data-processing manager, and sometimes to other managers outside the installation. Such reporting will emphasize the major structure of the system and suppress minor detail. He must report *downwards* to his programming support staff. This communication must be excessively detailed to make sure that all the system requirements are met by finished programs. The less detailed upward communications are discussed in chapter 8. The analyst-programmer communications, which involve the greatest degree of standardization, are considered in this chapter and in chapters 3-6.

Data-processing standards are the rules that govern the preparation of all work carried out by the electronic data-processing (EDP) department. They specify the content and layout of all documentation, and the procedures used within the department for carrying out all work.

There are four main reasons for standardization in and around a computer installation.

1. *To improve communication between the EDP department and user departments:*

    The EDP department is a specialist group that has many facets not easily understood by the layman, who is often unfamiliar with the technical language. Confusion of terms often exists between EDP groups themselves. The technical-language problem can be solved by the use of glossaries, but, without other controls, written communications between EDP and external departments will not be satisfactory. Written communications, such as management reports, control sheets, computer input request forms, and computer output should have a standard format. Where possible, standard preprinted forms should be provided.

2. *To assist communication within the EDP department itself:*

    A typical computer installation consists of a manager, systems analysts, programmers, control clerks, and data-preparation and operational staff. These groups will need to communicate with each other before, during, and after the implementation of the computer system. Standards prevent ambiguity in communication if they specify content, layout, and procedures as mentioned above.

3. *To control the efficiency of the EDP installation:*

    Management has the difficult task of controlling an EDP department's specialized functions in the same way

that it controls other departments. Using standard procedures and documentation insures a specified quality of
work production. Application of standards enables management to assess work in progress and to assign precise
responsibilities to all personnel.

Planning is an essential part of the EDP department.
For example, the data-processing manager requires schedules for computer time. Planning requires performance
standards both for people and machines.

Computer manufacturers will supply performance figures for their machines, but performance standards for
EDP personnel are difficult to create and implement. They
must be thoroughly tried, tested, and amended. They
must be continually checked and revised to maintain
their effectiveness. Performance standards cannot be created without standardized documentation and procedures.

4. *To document results from systems design:*

The analyst's aim is to produce documentation that will
finally result in a specific computer program. If the installation consists of a lead programmer, systems analysts,
and programmers, the analyst may present his documentation to the lead programmer, who will then prepare
program specifications for other programmers. On the
other hand, the documentation may be passed directly
to programmers. If so, it must be more detailed. Each installation should adopt a standard procedure.

2.6   Fact recording

The analyst may often be confronted by a complex existing
clerical system. He can ignore the existing system and design
a new one, or he can investigate the existing system thoroughly.
However, the first choice implies that the analyst is exception-
ally talented or that the original system has no merit, and the
people who operate the system will resent either of these
implications.

In nearly every case, the analyst must study the existing
system to understand its functions. Some of these functions

are essential and must be included in the computerized system. Since he does not know which are essential, he must study and understand them all.

He should select an unambiguous, clearly understandable method of recording facts about the present system that identifies facts to be cross-checked. The analyst should also obtain samples of all documents used in the system. Wherever possible, he should examine used copies instead of blanks, because many of the preprinted spaces on forms may not actually be used. If only blank samples are available, typical entries should be copied from a completed document. He should note any limitations on document size or layout.

Two rather specialized techniques for recording are tape recording and filming. Tape recording provides a lasting, accurate record of an interview, but it tends to prevent free interchange during an interview, and some people are reluctant to have their replies taped. Filming can be very useful in a systems investigation with a high work-study content, in which personnel movement has to be recorded, but it is expensive.

In the normal course of his work, the analyst is much more likely to use some charting device for recording facts. Organization charts, activity charts, and system flowcharts are three common types of charts.

Organization charts (see Figure 2-B) indicate the mechanistic structure of a department. They show the line functions and indicate the responsibilities of the personnel involved. Their value to the analyst lies in their display of levels of authority. They can give him an overall impression of the scope of the department under investigation from a well-drawn organization chart before he visits the department.

An activity chart describes a department's structure by specifying what each person does. It can provide the analyst with an indication of the man-hours involved in the throughput of work. A typical example is shown in Figure 2-C.

A system flowchart charts the flow of information through a specific procedure, or process. It is not concerned with the responsibilities of the individual performing the process, but only with the procedure itself. The analyst may be in unfamiliar territory. He may never have seen a certain process. The jargon of the system may be a language full of "pink slips," "bar numbers," "stop cards," or "Joe's book." A motorist in an

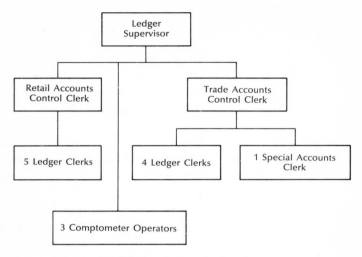

Fig. 2-B. Sample organization chart

unfamiliar country does not rely on a series of written or verbal directions; he gets a road map.

The analyst cannot buy a map: he has to make one. In doing so he learns the feature of his territory. An incomplete road map shows roads ending for no apparent reasons. An incomplete system flowchart shows similar gaps in clerical routines. The message to the mapmaker is the same in both cases: "Go and have another look."

System flowcharting has three main functions:

1. It enables the systems analyst to be reasonably sure that he has covered all aspects of the system.
2. It provides the basis for writing a clear and logical report.
3. It is a means of establishing communication with the people who will eventually operate the new system, since the user can be involved at this stage. He will be interested in seeing his own narrative appear on a chart in symbolic form and probably even more interested to find that he can understand the technique.

The purpose of a flowchart is to reduce a procedure to its basic component parts and to emphasize their logical relationships. A connected pattern of activity can thus be traced

ACTIVITY	PERSONNEL				
	Supervisor	Analyst	Analyst	Clerk	
Inventory control record and report (33 man-hours)			Sample check postings (8 man-hours) Prepare inventory report (12 man-hours)	Post orders (10 man-hours) File posted papers (3 man-hours)	
Requisition materials (23 man-hours)	Review and sign (6 man-hours)	Determine and separate cards requiring orders (6 man-hours) Make pencil copy of requisitions (8 man-hours) Maintain requisition control register (3 man-hours)			
Expedite shortages (20 man-hours)			Prepare and dictate expediting correspondence (10 man-hours)	Prepare and dictate expediting correspondence (10 man-hours)	

Fig. 2-C. Sample activity chart

and will highlight duplications and repetitive activities. Symbols used in systems flowcharts are shown in Figure 2-D, and flowcharting will be discussed in greater detail in chapter 6.

When a new procedure is designed, its flowchart can be compared immediately to the old procedure. Since flowchart symbols are standardized, anyone familiar with their meanings can quickly comprehend the documented procedure and its internal operations.

Figure 2-E is a sample flowchart representing the following procedure: The computer reads the customer order cards (from phase 1) and checks with the customer master file for credit information and other data. The product file supplies the computer with product data for the invoice and the shipping order. With this product and customer information, the computer creates order records and makes up an order file.

At this point in the flowchart, a new symbol is introduced:

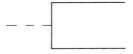

It is called a comment or annotation symbol, and it provides a place to write any needed explanation or clarification. The broken line is attached to the symbol or flowline being clarified. As the comment symbol explains, the computer first sorts the orders by order number and then within order number by warehouse shelf location. (This enables the order picker to pick the items for a given order most efficiently.) The computer then prints the shipping orders using information from the order file, and the shipping order is sent to the warehouse.

The page describing the action at the warehouse is not shown here; we can assume that the shipping order is received, the order is picked and shipped, and a card for each item shipped is sent back to the computer. Notice that there is a time lapse shown at the start of phase 3, indicated on the flowchart by a wavy line. The time lapse covers the warehouse and shipping activities that must take place before the notification cards are sent back to the computer. When the shipping department notifies the computer via card that shipment is made, then the invoice is printed and an invoice summary record is added to the invoice summary file.

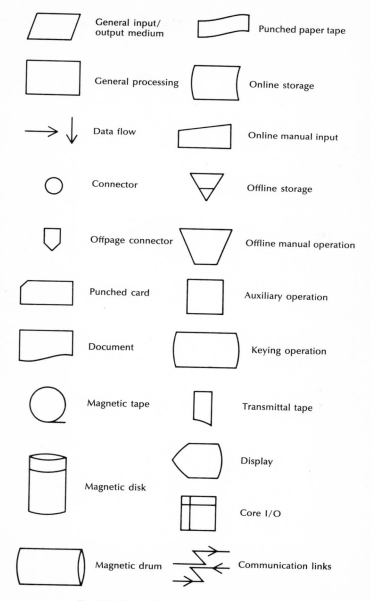

Fig. 2-D. System flowchart symbols (USASI)

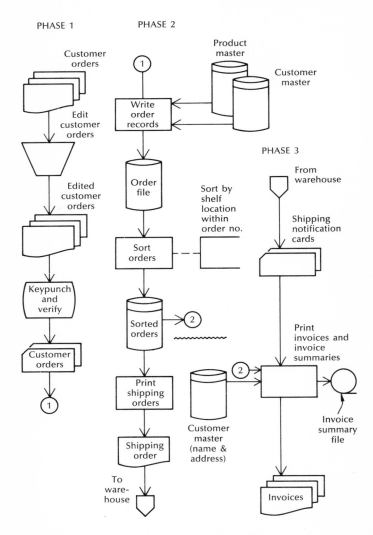

Fig. 2-E. Sample system flowchart

Exercises

1. What are the two stages of a systems analyst's investigation? Explain the difference between them.

2. How does a management audit differ from a normal pre-design analysis?

3. What general categories of questions should be included in a management audit checklist for a payroll department?

4. Explain the two main objectives of an interview in systems work. What are the disadvantages of this method?

5. When are questionnaires appropriate as a fact-finding method?

6. When are observation and sampling appropriate as a fact-finding method?

7. What are the main reasons for standardization in a computer installation?

8. What techniques are commonly used to record and organize the facts gathered during a survey? Which is best?

9. Explain the functions and advantages of systems flow-charting.

10. Write a systems flowchart showing the path a personal check takes from the time it is written until it is filed away after reconciling it with the bank statement.

11. Interview someone who performs a clerical job in an organization; determine what output in the form of reports or partial processing of data this person produces. Question particularly any apparent inefficiencies in his procedures, and ask him if he has ever suggested a better way to do his job. Note whether his reactions be-

come defensive, even though he knows you pose no threat to his security. How do you think this person would respond if you were a systems analyst asking these questions in order to design a new system?

input design

Input design occupies a large proportion of the systems analyst's time. Until the raw data of the new system can be organized in a form suitable for computer input, no processing can be performed, and no results can be produced.

The analyst must act as a link between the user of the system and its technicalities. Since people are involved, system requirements must be carefully balanced against personal capabilities. Conflicts are inevitable, and skill is required to resolve them amicably and yet achieve desired results.

This chapter discusses some of the problems of input design and suggests ways in which the task can be performed effectively.

Determination of system inputs 3.1

Inputs to the computer system can be either batched records or real-time entries from a terminal. Even if real-time entries are used, batch processing may be desirable. It can be achieved by the use of a terminal that transcribes the data to a suitable medium such as punched paper tape. Records on the tape can then be sorted and processed in batches at a convenient time. Regardless of the method of capture, the analyst must determine whether his input is relevant to more than one system and design it accordingly. For example, an order from a customer might affect invoicing, sales analysis, inventory control, production scheduling, and distribution systems, each of which requires different information. The transaction may need to carry more detail than is required for any one of its uses, or it may have to refer to a previous transaction. The analyst may need to consider whether the transaction should be processed before distribution or distributed in its original form. He may have to determine the extent of integration, which is essentially the technique of making multiple use of input data and using the output of one subsystem as the input to another.

Usually, the more integrated a system becomes, the more efficient is the use of the computer. However, the systems design required by an integrated system is far more complex than the design of one that treats each subsystem independently. A usual compromise is to plan an integrated system but to write each subsystem with unique input/output. When all the subsystems have been successfully implemented, the input and output sections can be modified and integrated in stages. This approach gives users more confidence than attempting to design an ambitious integrated system whose subsystems will not operate compatibly.

There are three main subdivisions of computer system input. *External data,* such as hours worked, quantities sold, and part numbers, is the prime input and the main reason for the system's existence. *Internal user data,* such as a request for special action or an interrogation, enables the user to communicate with the system. *Internal system data,* such as job-control parameters and console typewriter instructions, has a limited scope but is vital to the physical operation of the system.

External data consists of current data and file data. Current data is usually the major load in any system and typically is a large number of small items. File data, on the other hand, is often a small number of relatively large items.

The relation between the current data and its associated files is another problem that concerns the analyst. He must specify how data is to be handled. It is generally best to update all file data as soon as possible, particularly with file-maintenance input such as insertions, deletions, and changes in existing records. When batch-processing methods are used, it is usual to insist that file-maintenance data be submitted as a separate batch, which can be handled before current transactions, thus insuring that the file will be as up to date as possible. Sometimes run considerations in an installation determine that file-maintenance runs should be performed at regular intervals. It may be wise to submit insertions daily and store them at the main file until the regular update run is processed to minimize the possibility of error caused by the system's rejection of transactions affecting the inserted records.

When determining the inputs to a system, the analyst must recognize the important characteristics of an input item. The specification for each type of input to the system must detail —

1. Identification
2. Content and format, in specific terms, including recording media used
3. Frequency of receipt
4. Expected volumes, in suitable units, such as documents or characters
5. Conditions that govern its appearance in the system
6. Sequence in which it will be received for processing
7. Validation procedures

The analyst must also choose the input media to be used in the system. A number of points must be considered, such as the basic method of input (random or batch processing), what records of input are required, what peripheral devices can be obtained, whether the system is integrated or discrete, and the effective costs. The next section considers several methods of capturing input to resolve these considerations.

In some circumstances interactive conversational input procedures using keyboards to enter data directly into real-time systems are possible. Systems of this kind can be expected to proliferate. The design of the system's responses is a complex task and usually involves a great deal of development work.

Generalized time-sharing packages will probably be the major line of development in this area. The user will supply only the validation parameters, and the packages will then modify themselves accordingly.

Clerical procedures are another input method. The analyst should realize that the design of his computer input system will affect the design of the clerical system. He must distinguish between the "object" clerical functions directed to the main objectives of the system and the "systems" tasks associated with controls and other administrative procedures.

3.2   Data capture and preparation

Input methods must be reconciled with the user company's operating requirements. It is of little use to expect a salesman to complete a complicated form in a customer's office or to expect factory workers to comply with sophisticated and complicated methods of production control. Providing input may be the only contact with the computer department that most employees in a company have. If they regard this task as tedious and unrewarding the quality of input will suffer. Simplicity and ease of data capture and collection must be reconciled with the need to present data to the computer in an acceptable format. Accordingly, media used for other purposes as well as for computer input (dual-purpose media) are theoretically the most suitable method of data provision since they do not require data collection to be a special task. It is frequently possible to modify paperwork methods before the system is installed to conform to those the EDP system requires, so that the computer's arrival does not create an additional administrative headache.

Depending on the method of data capture and preparation, some or all of the following activities will be required:

1. Original recording, which documents the original transaction for subsequent processing
2. Transmission, which moves the data from the point of origin to the data-processing center
3. Transcription, which records the original transaction on a suitable input medium
4. Verification, which insures that the transcription was performed accurately

5. Sorting, which orders the input into a sequence acceptable to the computer program
6. Control, which validates that all original records have been transcribed
7. Input, which is the system's assimilation of data after preparation and before processing

In principle, the fewer stages the better, for reasons of time and accuracy. In practice, however, fewer stages cause more expense.

Ideally, data should be recorded in a machine-processable form as a by-product of the actual transaction or event. However, this can be both expensive and unsatisfactory. Economy and availability of trained staff, machines, and servicing capabilities frequently require centralized data preparation. If centralization is to be adopted, the analyst must choose an appropriate method of recording and transmitting data. This depends on the application involved, the overall system timing requirement, the volumes of data to be processed, and the cost of equipment in terms of its benefits. The following methods are suitable for operation on either a centralized or remote basis.

1. Card punches automatically produce 80-column cards and can also produce 21-, 40-, and 90-column cards on suitable equipment. The Port-a-Punch is a small manual instrument using prescored cards that can be punched by hand with a stylus. It has a very limited use, and, since it is completely manual and unverified, it must usually be subjected to extensive pre-editing and reasonableness checks.
2. Paper tape punches can output 5-, 6-, 7-, or 8-channel punched paper tape with various code patterns, depending on the manufacturer.
3. Magnetic tape encoders are capable of writing data directly on magnetic tape. This method is expensive, but it is likely to become more widespread.
4. Indirect keyboard machines produce punched cards or paper tape as a by-product of a normal keyboard operation to prepare hard copy. Examples are typewriters, accounting machines, and cash registers.
5. Mark sensing readers/punches automatically convert

pencil marks made in predetermined positions on a card into punched holes on the same card, thus eliminating manual punching of data.

6. Kimball tags are small perforated tickets, commonly found on stock in ladies' dress shops. When fed into an automatic converter, they produce standard punched cards for computer input.

7. Magnetic ink character readers translate the stylized fonts printed in magnetic ink on documents (particularly checks) into direct computer input.

8. Optical character readers are very similar to magnetic ink character readers, except that they can read pencil or ink. Characters are recognized by their shapes instead of by a magnetic pattern. These devices can be used as free-standing units to produce cards or paper tape, or as direct input media to the computer.

9. Mark readers can also be free-standing or direct input and resemble mark sensing readers/punches, but use normal sized paper forms instead of punched cards.

10. Remote input units automatically create input data when fed a plastic badge or a prepunched card. Variable (numeric) data is inserted from the same unit by keyboard.

11. Online terminals often resemble a typewriter and allow formalized messages to be typed to interrogate a file. The computer relays the answer to the input unit via a screen, or types it via the online inquiry console.

The systems analyst must determine which method is most appropriate. If involved in the selection of equipment, he must decide which is the most suitable and economic in terms of the company's operating requirements. It may not always be valid to reject a method of input because it seems at first to be unconventional or expensive. For example, in certain circumstances, an optical character reader could profitably replace four punches and three verifiers.

One set of factors—type of processing, speed, accuracy, verification, rejection rate, operator requirements, cost, and relevance—is common to all forms of input. By evaluating them, the systems analyst can determine whether a particular method is suitable for a particular application.

The type of processing may be batch (groups of records) or online (individual and distinct records, processed as such), sequential (sorted) or random (unsorted). For example, punched paper tape may be unsuitable for sequential processing because it cannot be sorted offline. Punched cards are generally unsuitable for online processing because considerable preparation time is required.

The analyst must also consider speeds of capture, preparation, and entry of data into the computer. For example, punched cards and paper tape are slow to prepare, but are fast in entry when compared to some other methods. OCR documents need little preparation, but are read slowly. If transcription can be avoided, input may be more accurate. If a large volume of data has to be accumulated (for example, for a statistical report), a certain level of inaccuracy may be acceptable for reasons of economy. Several methods of verification are available. They include sight checking from hard copy, arithmetic methods (check-digit verification, for example), re-entry on punched cards, automatic program verification, and batch-level verification.

The rejection rate may be estimated by anticipating that a certain number of documents will be unsuitable for automatic handling. Operator requirements may be determined by estimating the frequency of need for human intervention.

Cost considerations must include the capital cost of equipment as well as the cost per character of input. Input/output media, floor space, staff, and overhead costs must be included in the latter. Relevance involves considering a device's inherent suitability for a specific application (for example, Kimball tags hang well on clothes, and a MICR reads checks well). Other factors that the analyst must also take into account are the relevance of optional devices, possibilities for expansion, and suitability for other applications.

Table 3-A summarizes the various methods of data capture and indicates their relative performance. Obviously, suitability for specific applications cannot be shown on a table of this type, but must be judged by the systems analyst.

The two principal input media are punched cards and paper tape. Each can be used advantageously in certain applications. Partial lists of advantages for each medium are given below.

TABLE 3-A
COMPARISON OF DATA-CAPTURE METHODS

| METHOD (AND VARIANTS) | TYPE | SPEED | | ACCURACY | VERIFICATION | REJECTION RATE | OTHER FACTORS |
		Capture and Preparation	Input				
Punched cards	Batch sequential	4k–12k cph per hour	400–2,000 cpm	Good	Re-entry	Low	Plastic cards
Prepunched	Batch sequential	Pull 300–500 cph	400–2,000 cpm	Fair	Automatic	Low	Round-cornered stub cards
Mark-sensed	Batch sequential	1,800 cph	400–2,000 cpm	Fair	None	High	
As a by-product	Batch sequential	Automatic	400–2,000 cpm	Good	Automatic	Low	
Port-a-punch	Batch sequential	1,200 cph	400–2,000 cpm	Fair	None	Fair	
From Kimball tags	Batch sequential	Automatic	400–2,000 cpm	Good	Automatic	Low	
Punched paper tape	Batch sequential	6k–12k cph	300–2,000 cps	Good	Re-entry	Low	5, 6, 7 or 8 channel

	Batch sequential or random	Approx 9k cph	9k–340k per sec	Good	Re-entry	Low	Restricted choice of equipment
Magnetic tape	Batch sequential or random						
OCR devices	Batch sequential or random		200–300 dpm	Fair	None	High	Font printing
Bar codes	Batch sequential or random		200–300 dpm	Fair	None.	High	
By-product tally roll	Batch sequential or random	Automatic	200–300 dpm	Fair	None	Fair	
MICR devices	Batch sequential or random	Automatic or keyboard	1,200–1,600 dpm	Good	None	Fair	Sorter/readers
Remote terminals	Online random or batch	100 cps		Good	None	Low	Card or punched paper tape output
Online terminals	Online random	Demand fast		Good	None	Low	Interrogation badge-reading
Analog to digital conversion	Online random	Demand fast		Good	Automatic	Low	Online control

Some advantages of punched cards are—

1. They can be sorted and processed offline by punched card equipment.
2. They are interpreted and read by eye and are therefore a valuable and permanent visual record.
3. Their usual fixed format reduces error possibilities.
4. They allow easier correction of errors (incorrect data is more common than punching errors).
5. They can be prepunched and used as direct input.
6. They can be a dual-purpose medium.
7. They are considered by some to be sturdier than paper tape.
8. They are easier to use for programming input.
9. Their files are neater than tape.
10. They can be loaded on a file-feed device while the computer is operating, whereas time may be lost in changing reels or waiting for tape rewinds.
11. They allow individual records to be easily inserted and changed.
12. The standard 80-column card is universally accepted; there are no conversion problems caused by a wide range of formats as in 5-, 6-, 7-, and 8-channel tapes.

Some advantages of paper tape are—

1. It is easier and cheaper to produce as a by-product to the production of hard copy.
2. It prevents sequenced records from getting out of sequence, as they can in card files.
3. It is less bulky and contains more information per unit weight, and therefore is cheaper to store and mail.
4. It accommodates variable-length records easily.
5. It provides for parity checking.
6. It is compatible with Telex and generally is more suitable than cards for data transmission.
7. It has both upper- and lower-case facility.
8. It is cheaper, unless card-image techniques are used.
9. Paper tape readers and punches are cheaper than card equipment.
10. It can be read in reverse.
11. It may give a faster character reading rate.

12. A paper tape reel is easier to feed to the computer than its card equivalent.
13. Errors during punching can be designated and ignored in processing; whole records do not have to be re-punched.
14. It has a lower redundancy rate, and blank fields do not have to be read as do unfilled areas of a card.
15. There is no need to zero-fill or to repeat control information as is necessary when a record extends over more than one card.

Forms design 3.3

The systems analyst may have to design many types of forms. Some of them may be for parts of the revised system that are not directly connected with the computer. Therefore, the analyst must be aware of the general principles of good forms design. He cannot restrict his knowledge to special considerations that apply only to the design of punching documents.

There is no simple guide to forms design. The subject includes a mass of detailed information in which it is easy to lose general principles. The analyst will learn to design good forms through practice, and if his company has forms design specialists, he should enlist their cooperation.

Forms design is often assumed to be simply the positioning of information on a blank piece of paper in a logical format. Although this aspect of design is important, it is neither the only nor the primary consideration. Many defects in forms result from insufficient attention to how the information is collected and used. The completed form, as a whole, must be suited to the conditions in which it is intended to operate.

Factories, department stores, warehouses, offices, computer rooms, building sites, and distribution centers all use forms, but the conditions under which they are filled out differ. In some cases, machines or clerks at desks do the work; at other extremes, workers in shipyards may count numbers of rivets and record them on worksheets resting on their knees while perched on a scaffold in the rain.

The forms designer can make the task of completing forms much easier if he provides the form on suitable material. Light, thin papers may be suitable for situations were accou...ting machines are used, but these papers disintegrate rapidly when exposed to weather. A wide range of materials is available.

Whether ballpoint pen, pencil, typewriter, or tabulator is used to complete a form depends on where it is used. It may be necessary to provide backing sheets for documents. A medium that will not smudge when damp or dirty may have to be chosen. A paper heavy enough to bear the pressure of a ballpoint pen may have to be used, or a lightweight paper may be chosen for a tabulator.

The size of paper used for a form ultimately depends upon its contents and use. It should be within the usual standard paper sizes wherever possible, which not only keeps the cost of printing down, but also insures that it will fit in standard files and binders.

The form should not be so large or so small that it is unmanageable when filling in, punching, binding, filing, or perforating. A related problem arises when the document has to be inserted in an envelope. Problems may also arise when folding for window envelopes. A significant amount of time is saved when window envelopes are used and addresses need not be retyped. Where folding is critical, folding marks should be arranged so that an edge is folded to a particular mark. A folding machine should be considered for handling large volumes of forms. This in turn may affect the design. Also, it may be necessary to specify the type of envelope required for a particular form. All envelope specifications should be checked with the post office.

The legibility of a form depends on the medium used to complete it, the paper, and the number of copies. Certain forms must be extremely legible to avoid error and to increase speed when they are used as punching documents. Other forms that are simply filed as records need not be as legible.

Color sometimes has to be considered when designing a form. Colored paper or colored printing on white paper is used to differentiate between copies and to assist in correct routing. This use applies particularly to multi-part sets. Another valid use of color is to highlight a key area or to segregate entries for electronic sorting. The legibility ratings devised by Le Courier for various color combinations is given in Table 3-B. Color-blind persons may confuse light blue with pale green, orange with pink, buff with green, and red with brown.

When using carbons in a multi-part set, it is advisable to experiment with various grades of carbon paper. Some carbons are reusable but give poor copies if more than three or four are required. Other carbons give acceptable copies throughout a set, but cannot be reused extensively. NCR paper is an alternative to be considered. It is chemically treated to produce copies when pressure is applied. Disadvantages of NCR paper are its cost and its tendency to mark easily, particularly at the edges.

TABLE 3-B

LE COURIER'S LEGIBILITY TABLE

Order of Legibility	Color of Printing	Color of Background
1	Black	Yellow
2	Green	White
3	Red	White
4	Blue	White
5	White	Blue
6	Black	White
7	Yellow	Black
8	White	Red
9	White	Green
10	White	Black
11	Red	Yellow
12	Green	Red
13	Red	Green

The number of copies in a set of forms should be kept to a minimum. The more parts to a set, the greater the cost. There is little merit in having an extra copy as a safeguard; it causes confusion during implementation and during running of the system. If another copy becomes necessary after printing, it can be included when the form is reordered. The first batch of forms should be ordered in an economic quantity; however, the quantity should not normally exceed four months' supply, so that supplies will not be wasted if changes become necessary after practical use.

There may be a need for a form or part of a set to be photocopied or duplicated. Certain aids to good reproduction are available, depending on the method of reproduction to be employed. They can be incorporated at the form design stage.

If the finished form is to be printed outside the company, the analyst must discuss most of the technical points raised above with the printer's representative. Certain printing operations can be more expensive than expected, and it is wise to guard against them during the drafting stage.

The printing profession uses many specialized symbols to indicate sizes, positions, and alterations. These symbols are often listed as though for general use, but it is usually dangerous for the layman to use them. A scale drawing of the form with the required dimensions is satisfactory. Samples of the type face and the form of lines required should be attached to the drawing. The weight, type, and colors of paper should also be specified or samples should be attached. The printer's proof must be carefully scrutinized against the specified design, since subsequent changes will be very expensive.

Sequential numbering is done with a consecutive numbering machine. It normally consists of 4, 6, or 8 digits and has limited facilities for using alphabetic characters. Using elaborate sequential prefixes and suffixes may require special arrangements at considerable added expense. Also, special requirements such as collating and inserting carbons in edge-gummed sets, patch carboning, gumming, drilling, and perforating can also increase printing time and expense.

If a form is a single document that will be used for only a short time, it can often be most conveniently and efficiently produced on in-house machinery. Assuming that the printing equipment exists, the quality of output required and the length of the run (quantity) will determine the method of producing the form. Approximate values for the relative quality, run length, and cost of various methods are shown in Table 3-C. Generally, there is a definite correlation between run length, quality, and speed of the reproduction process. A list of precedence for methods of reproducing forms is shown in the last column of the table. From the table, we can see that offset printing is firmly established in the top three positions. Paper and electrostatic offset plates are easy to make and run on simple automatic machines. Direct electrostatic copies are second from last because of high cost and low speed. They are important primarily as a convenient means of producing a small number of copies. The table does not allow for the effect of the quantity discount rental plans

TABLE 3-C
Cost Comparisons for Internally Printed Forms

Method of Reproduction	Rankings						Cost/Run/ Quality Precedence
	In Order of Quality	In Length of Run Obtainable	In Cost Per Copy				
			Number of copies				
			10	100	500	1,000	
Offset (foil/metal plate)	1	1	5	4	2	2	3
Offset (paper typed plate)	2	2	2	1	1	1	1
Electrostatic Copies	3	8	4	5	3	3	7
Offset (from electro-static paper plate)	4	3	3	2	2	2	2
Wax Stencil (typed/pencut)	5	4	2	2	2	—	4
Spirit (typed/drawn master)	6	6	4	3	—	—	5
Spirit (heat transfer master)	7	7	1	1	—	—	6
Wax Stencil (heat transfer)	8	5	1	1	—	—	8
	1=best	1=longest	1=cheapest				

now offered by some manufacturers. Spirit copies are low on the list because of their unsuitability for long runs; for short runs they are effective and inexpensive.

The final design of a document is a compromise between ease of completion and ease of processing. The relative priority of these two factors should be carefully assessed, since it is rarely possible to design a document that is ideal for all purposes. The establishment of priorities inevitably helps decide the design. When cards or paper tape are to be punched from a document, there is little doubt that convenience for punching (or processing) is of far greater significance than convenience in filling out the form. On the other hand, the completion of invoices by traveling salesmen or the maintenance of inventory by a storekeeper requires simple documents designed for rapid and accurate completion.

Once priorities have been determined, the analyst should prepare a list of the information that the form must carry. If ease of completion has priority, the list should be compiled logically in the order in which the information becomes available. If processing or analysis of information has priority, the checklist should consist of the data in the logical order used, followed by information rarely referred to after the form is completed. Figure 3-D shows an example of a checklist prepared in designing a card for manual inventory recording.

Once a logical checklist has been designed, columns and block areas should be sketched out according to the sizes required. The analyst should decide the most suitable number of lines of variable information. For example, if an average of 3 items are sold per invoice, there is little point in allowing more than 10 lines to cover all exceptions.

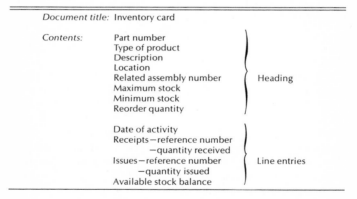

Document title: Inventory card

Contents:
Part number
Type of product
Description
Location
Related assembly number   } Heading
Maximum stock
Minimum stock
Reorder quantity

Date of activity
Receipts—reference number
            —quantity received
Issues—reference number   } Line entries
            —quantity issued
Available stock balance

Fig. 3-D. Sample form design checklist

The analyst should clearly indicate margins on the form. It is difficult to write or type on the top and bottom half-inch of a document. He may allow the top half-inch to contain part of the title, and the bottom half-inch may show each copy's destination. Space at the left is usually required for binding or filing. If the form is numbered sequentially, the number should be in the top right corner.

His next step is to draw the outline of a form of the appropriate size on a forms design chart, squared with measurements to suit the spacing required. If such paper is not available, scale spacings can be marked along the edges of the form as a guide. Any area that will be required for punching, binding, or filing should be clearly marked. If ease of completion has priority, he should follow the checklist and work down the form from the top left, inserting areas for the information required. For a symmetrical layout, areas of similar size should be grouped together as far as the checklist will allow. However, some information may have to be placed elsewhere because of other considerations. Addresses are usually in the top left; sequential numbers in the top right. If window envelopes are used, address positions will be affected. Whenever areas are predetermined, as in this case, they should be allocated first, and remaining areas arranged around them.

If spacing depends on the use of a typewriter or tabulator, design is relatively simple. Vertical spacing is usually based upon multiples of 6 lines per inch. Specifications are usually given with the machine, and a simple test measurement will confirm the spacing to be used. Horizontal spacing is usually based on units of either 8, 10, or 12 characters per inch. Drafting can be simplified by using a print chart.

The analyst should keep in mind that handwriting is usually larger than typewriter or tabulator print, but he should not allow too much space. If space is limited, the writer tends to write more carefully and legibly. When a form is to be completed by hand printing, a series of boxes for check marks encourages uniformity and proper spacing; this is especially important when the forms are to be processed on optical character reading equipment. A safe criterion for handwritten forms is to allow eight characters to the inch. For mechanical completion of forms, allow space for one more character than the maximum required to help the typist or computer operator align the form horizontally.

Thin straight lines should surround each area of information. For columns of figures, sets of three lines at normal spacing and then one line at double spacing help the eye follow the figures. The same effect is obtained with less space wastage by two dotted lines and then one solid line, throughout the length of the form. Continuity of lines simplifies typesetting and promotes the impression of planned design. However, it is rarely necessary to put lines around the edge of the form.

Generally, the information specifically identifying the form should be put in the top left corner. This area should be clearly marked with space for whatever information is required and should not be larger than required for the information to be entered. Boxes for check marks are particularly useful on forms containing a series of possible responses.

In multi-part sets, obliterating information on a part of the set is often necessary. Usually, such information is not to be conveyed to certain persons (for example, cost prices to customers). It may be possible, by careful design, to avoid putting this information on some copies of the form. By using shorter paper for those copies that do not require the information, the problem can be solved simply and inexpensively. Spot carbons may also be used effectively, but are relatively expensive. Another method is to interleave a plain piece of paper to cover the column not required. After the form is completed, the paper can be withdrawn and destroyed.

Having completed a detailed first draft, the analyst should discuss it with another analyst, if possible, and with all who will use the form. The design should be changed if necessary. He should submit the final draft to the user department's manager for formal acceptance. If possible, his signature should be obtained so that the expense of any subsequent alterations would be his responsibility.

The systems analyst has a particular responsibility for the design of forms that will carry data to be punched on cards or paper tape for input to the new system. In the case of punching documents, ease in processing should be the prime consideration because of the need for fast, accurate punching. If it is impossible to design a source document that can be both punched and completed easily, then transcription may be unavoidable, but is expensive and introduces errors. If it is possible, the information checklist must be designed in conjunction with the field layout of the card or paper tape to be punched.

Information should be laid out from left to right in the order of punching. If all information cannot be arranged in one line, lines of data should not be placed too closely together. In some cases, it may be preferable to present the data arranged item by item vertically. Fields should be separated by bold lines to help the punch operator keep his place.

Keypunch operators may prefer to punch from certain areas of a document. When the operators' standard practice has been to punch from the top, the middle, or the bottom of a document, the practice should be continued if possible. When no such precedent exists, perhaps the best area to place information for punching is the lower part of the document. If this is done, the start of the punching area and the end of other information can be indicated by a bold line. The operator, working left to right, will then punch everything below that line. Wherever practicable, the information for one record should be on one side of a single page. If two or more separate punching operations are performed from the same document, the area for the first operation can be boldly lined, and the second can be colored or lightly hatched. As many records as possible should be punched from one page to minimize page turning and encourage a smooth working rhythm. Information common to a group of cards should be presented first, so that it can be gangpunched.

Indicating the card columns in which data fields are to be punched helps in both punching and error correction. Shading is useful to divert attention, to avoid misplacing of information, or to obliterate unwanted information. Boxes can be used to focus the keypunch operator's attention. They can also insure that information is placed in correct positions, which is especially important when recording decimal data.

Codes can often be used with boxes on punching documents to convey more information on one form. Coding can be entered from a separate list or printed on the form itself. Examples of types of coding are shown in Figure 3-E. Coding can be preprinted fixed information, as in example 1, or it can be entered from elsewhere, as in example 2. If the number of choices is small, the code can be selected when completing the form (example3). If there are many items to choose from, the choice can be indicated by a check mark to avoid transcription errors. The keypunch operator can then select the corresponding item (example 4).

When another form may be required in the system, the following questions are relevant:

1. What are the purposes of the form? Are they important? Does the form accomplish them?

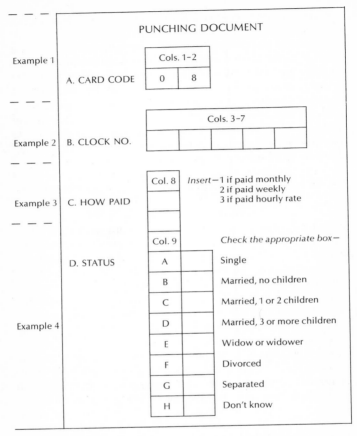

Fig. 3-E. Sample coding on punching documents

2. Can an existing form be used?
3. Can this form be combined with some other form?
4. Should the form be divided into separate forms?
5. Are all necessary copies included?
6. Are all copies necessary?
7. Does the title indicate the use of the form?
8. Is the title distinctive?

When drawing up the checklist of the tentative contents of the new form, more questions arise.

1. Has the form control number been included?
2. Is all repetitive data preprinted?
3. Have spaces been reserved for all required data?
4. For forms to be transmitted, should spaces be provided for "to" and "from" information?
5. Has space for signatures and approvals been included?
6. Should simple routing or handling instructions be printed on the form?
7. Is sequential numbering desirable?
8. Are all included items necessary?
9. Are all necessary items included?

When the new form has been justified and the contents are known, the analyst makes his first draft, and the following questions arise.

1. Have adequate spacing and margin standards been used?
2. Has the spacing been tested?
3. Is the most important data in the most prominent location?
4. Should both sides of the form be used?
5. Is data required for filing and sorting prominently located?
6. Is spacing sequence logical? Does it allow for minimum hand travel and ease of motion?
7. Is the sequence of items the same as on forms that are used with this one?
8. Is the sequence of items the same as on similar forms?
9. Should the form be designed for a window envelope?

When the form must be approved, the following questions should be considered.

1. Have all users and potential users been consulted for suggestions?
2. Have the persons responsible for the system, such as supervisors and systems men, been consulted for suggestions and approvals?

When the final draft of the approved form is prepared for the printer, the following questions should be checked.

1. Is the size standard?
2. Are the grades and weights standard and correct for usage, expected life, filing, and handling?

3. Has colored paper been considered?
4. Have all other required items such as size of order, method of printing, ink, holes for binders and special machines, collating requirements, and carbon requirements been accurately specified on the printing request?

3.4   Input validation

All data that enters the system must be correct; invalid input leads to incorrect output. Some error-control methods are built into the data-preparation equipment itself, but errors can still occur. For example, a keypunch operator and a verifier operator may both misread an input document. In this case, machine verification does not prevent an error on the input punched card. Similar errors can be introduced by the malfunctioning of the verification equipment. There is always a possibility that unverified data may enter the system.

In validating input data, control can be exercised at four main points:

1. Before input to the processing system
2. At the data-preparation stage
3. On input to the computer
4. Within the computer program that accepts the data

Data should be checked visually both for errors and to determine whether it is acceptable to the system before it enters the processing system. It can be read by the converting mechanism or by the keypunch operators. For example, if it is to be fed into a mechanical reader, it should be checked to see that documents are undamaged, the correct color, and so on. In many cases, data first enters the processing system at the data-preparation stage. At this point it should be checked to see that it corresponds with the data that was expected according to schedule. (The validation techniques on computer peripheral devices and as part of standard control software will be discussed in later chapters of the book.)

The main point at which input can be validated is within the computer program that accepts the data. The analyst must specify such a program with extreme care. The first question to be asked is whether the input is right for the process. The use of codes to identify types of data is the first step in good input control. The computer program can check to determine whether the proper code number is indicated.

After the type of data is determined to be correct, each item of information must be examined and checked. The analyst's function is to discover and record the standards against which input data is to be measured, and communicate them to the programmer. He, in turn, must be able to measure each item against these standards by its position, size, format, and perhaps by some delimiting signs, such as record marks. Care must be taken to distinguish clearly between fixed- and variable-length fields. For example, if the date is to be entered on a punching form, exactly how it will be punched must be specified. For example, if the data has a fixed format, the sixth of June must appear as 06/06/71. However, with a variable format, 6/6/71 may be valid. Although the fixed-format arrangement is easier to check, the problems of form completion must be considered. In the example above, using the fixed format requires that one remember to enter leading zeros when writing single-figure days and months.

When specifying a fixed-length field, indicating the total number of characters in the field is sufficient. For a variable-length field, the maximum and minimum numbers of characters that can be present must often be specified. The definition of a field should specify which types of data characters are valid, as a measurement standard. This is usually done by specifying A for alphabetic characters, N for numeric characters, and X for any characters within the available character set. Punctuation symbols, dollar signs, and so forth are normally written as literals. Hence, the format of a field that can contain ABC-1234 is specified as AAA-NNNN. Such a statement is often called a *picture* of the data.

Within a picture, it may be advisable to indicate the actual contents by specifying range checks. For example: "If the first character of code number is A, the range of numeric information is 1000 to 2500; otherwise, the numeric range is 0001 to 0999." An unambiguous statement of this type allows the programmer to test each piece of data as it is received. However, this kind of range specification may not be suitable for all types of numeric input data; other types of tests may be required. For example, if there are gaps within a sequence of valid numbers, comparing each number against a list is a more appropriate test. If the number contains a check digit (which is described more fully in section 3.6) its position must be stated clearly in the picture, and the programmer must be told explicitly how it is calculated. Another situation occurs if the number is inconsistent with related data. For example, in a distribution application, retail customers may have account numbers starting with numerals from 0 to 6 inclusive. Wholesale customers may have account numbers starting with 7, 8, or 9. Retailers may have discount terms A and B, and wholesalers may have discount terms C and D. If input range checks were applied to these two fields independently, the combination of customer 8324 with discount terms B would pass, although it is an obvious error. Also, a range specification would not be a valid check if a number violated some policy statement concerning invalid numbers. Examples of this type are serial numbers on stopped checks and account numbers for customers with bad debts outstanding.

In addition to examining each piece of data in isolation, control totals should be established before data enters the data-processing system. These can be totals useful for other operations, hash totals, or control totals from other systems; they will help detect invalid data before it is processed and insure that errors are discovered while source data is still available.

The objectives of a control system are to detect, locate, and correct errors. Error reports must be issued unless a self-regulating system is designed. They can be prepared immediately after an error is found, when all data has been checked, or when a logical batch of data has been checked. Turnaround time should be the criterion for selecting a reporting method, since a report is not required until action can be taken.

One advantage of computer processing is that error correction procedures can be monitored, and receipt of corrected data can be verified. In some cases facilities to bypass errors must be built into the system if unusual items of information may be correct in themselves but rejected by controls.

Data-preparation control   3.5

In the previous section, controls of input document data were described. They were designed to insure that the data had been transcribed correctly onto the input medium and had passed certain validity checks. However, input controls should also determine whether all the required data has been received and whether the data has been processed completely through the system. In order to determine this, the analyst must examine the data-preparation controls that are applied before data is converted to a form suitable for computer processing and again after computer processing has taken place.

Before any conversion is undertaken, whether all expected data have been received should be checked. Such a control system works best if input arrival can be predicted. For example, if orders for punching are expected to arrive each day and do not appear, they should be investigated.

After the expected documents arrive, they should be checked to make sure none have been lost in transit. This check can be made easily if data is submitted in batches accompanied by control slips containing the number of documents and some form of control total. (A control total may be generated as a by-product of other work, be specially generated for control purposes, or be a hash total designed simply to prevent losses in transit.) The submitting area can also number input documents sequentially and note the first and last numbers on the control slip. If documents are to be handled extensively within the data-preparation center it may be advisable to number them sequentially immediately upon receipt.

The advantages to be gained from batching incoming documents are:

1. Work can be readily allocated among various operators.
2. Locating errors is easier.
3. Each batch can be controlled individually.
4. An incorrect batch can be removed so that the remaining input can be processed.

Batch size must be calculated. It is affected by the number of operators and machines, machine and computer setup time, and a reasonable conversion time per document. If batch control totals are not submitted with input, it may be useful to create them before conversion begins. With some types of data-preparation equipment such as paper tape punches linked to accounting machines, control totals may be built during the conversion process. If punched cards are used as input, control totals may be obtained after conversion by adding relevant fields and summary punching their total value into control cards. If random input is accepted, each document is treated as a batch. Only overall controls can be accumulated in the computer and compared with manually accumulated totals.

Data for real-time systems is usually unsuitable for handling through a centralized data-preparation section. Data handling is usually reduced by having the user input data on-line via inquiry stations. In quick response systems, control can be applied by the anticipation method, in which an error condition is recognized if an answer is *not* received within a specified time. However, these systems usually involve data transmission that presents other control problems.

3.6 Code design and check digits

A systems analyst is rarely required to invent an extensive code system. Codes tend to be part of existing systems and to have grown with them. The need for designing a major code usually arises when a new system is superimposed on an existing organization; when a code is outgrown and cannot sensibly be extended, because to do so would destroy its structure; or when two organizations with different coding systems merge, and a single code is needed for both.

The function of a code is to make efficient identification or retrieval of coded items possible. This applies in normal data-processing activities as well as in information retrieval systems. In the latter, the information structure rather than the item listing is coded. The code fulfills this function by providing a substitute name for the normal item name. From an information point of view the name consists of a set of irrelevant characters, but it can do two things. It can reveal a relationship to other items of a similar kind, and it can reveal properties of the item itself. A normal item name does neither of these.

A data-processing code must fulfill the following functions:

1. It must be logically tailored to the system.
2. It must be so precise that any item can be described by only one code word.
3. It must use a minimum number of characters.
4. It must provide for all system expansion and development that can be foreseen.
5. It must have a clear structure, so that the user can understand it and encode items without error. Possible sources of ambiguity must be clearly explained.
6. It must use mnemonic aids if clerical or other non-computer reference is required.
7. It should suit the data-processing software and hardware, the system of storage levels, and the indexing systems with which it is to be used.

Systems considerations of code design raise the following questions:

1. What is the data to be used for, and in what sequence?
2. What is the function of the code? Who will use it?
3. What is the structure of the data and the file management system?
4. Are item lists or an information structure being coded?
5. What is the size of the file in terms of the hardware provided?
6. What is the rate and character of item searching?
7. What is the file's growth rate? If the data is hierarchically organized, are the main concepts or the low-level subsets likely to increase?
8. Will the code ever be used visually for clerical identification? If so, which parts of the code word will be involved?

Closely allied to coding is the need for classification, which can be defined as the systematic arrangement of all items within a system so that similar items are grouped. Within the resulting framework, individual members of a group are further defined according to their fundamental attributes. An acceptable coding system can emerge naturally from well-classified data.

If a coding system does not provide for minor changes, its basic system of classification is often deficient. An example occurs in sequence codes, where the next available number from a list is allocated to each item in turn. These simple codes are rarely used for representing more than 20 to 30 items, because of their inflexibility. It is impossible to insert new items into the sequence, the code contains no useful information about the items, and there is no correlation between the code and the items represented.

A better code is the group classification code in which all or some of the digits in each code number indicate a particular classification. For example —

          5xxx   Purchases
          51xx   Production material purchases
          511x   Steel purchases
          5111   Steel plates
          5112   Steel strips
          5113   Steel wire

Note that the last digit in this example is a sequence code.

When a numerical code is used, up to ten classifications are possible for each digit in the code. Each classification is represented by one digit from 0 to 9. In some codes, one or more digits are not used when the system is designed, but are reserved for new classifications that may arise.

When an alphabetic code is used, up to 26 classifications are possible for each position in the code. In practice, because of possible misinterpretation, it is common to use only 21 letters, which restricts the code to 21 classifications.

The attributes of items that are to be coded do not always fall into ten main groups, ten subgroups, ten subsubgroups and so on. If less than ten groups are to be classified, the code will not be fully used. In that case, a block code is an alternative. The block code is virtually the group classification code modified to provide more groups with less digits, but providing room for limited expansion. For example, if there are 4 main groups (A, B, C, and D) respectively consisting of 15, 21, 18, and 24 items, the normal group classification would require a three-digit code for these 78 total items. One possible solution using a two-digit block code is as follows.

Main Groups	Code Numbers Allocated	Spare Code Numbers in Each Group
A	01-19	4
B	20-46	6
C	47-69	5
D	70-99	6
	Total	21

Block coding systems of this kind are easily expandable within each category. They can, with practice, be associated with the items or groups they represent. Whole sections can be added or deleted. The user or computer can easily check code numbers as they are received. The code can often be used as a basis for sorting and limited information retrieval.

Another simple coding technique is the use of significant digits. Individual digits represent features of the coded item or are given some other special significance. In general, these codes contain actual facts about items. For example, TT 670 15 B may represent tube type, size 670 x 15, blackwall; and TT 710 15 W may represent tube type, size 710 x 15, whitewall.

Significant digits are often used for inventory items that require clerical reference. The code can be clearly related to the article that is retrieved and can be easily changed and extended. These codes are sometimes called faceted codes. Each of the component groups making up the code number may be thought of as describing a different facet of the individual item. When a faceted code has a free form, in which each facet can take any number of characters, the resulting numbers may be long and difficult to process. This disadvantage can be overcome by limiting the number of digits of each facet. This method has an added systems advantage. For example, consider the following code structure, which is designed to describe materials that will be used in some type of production.

1. Facet 1: Cross section (single digit)
   1 = Round
   2 = Square
   3 = Hexagonal

2. Facet 2: Material (two digits)
   01 = Brass
   02 = Steel
   03 = Stainless Steel
   04 = Wood

3. Facet 3: Length in inches (always 3 digits; if less than 10, prefix with 2 zeros; if less than 100, prefix with one.)

4. Facet 4: Type of finish (two digits)
   00 = In raw state
   01 = Cast
   02 = Machined
   03 = Planed

With such a code, a plain metal pin can be numbered 10204502. It is then uniquely identified, and the code can be used as a descriptor or key in a data-processing system. For example, this item now known as 10204502 might have previously been described as Part 123 (Pin), Part 394 (Dowel), Part 871 (Valve Roller) in the same store. Such occurrences are very common, and duplicate supplies of an item are often unrecognized because of inefficient coding.

A clothing manufacturer's code might have type, size, style, and cloth as its facets. The code for this application might, for a given garment, look like —

SU M 38L 17 384 Suit, male, size 38, long, style 17, material 384
SO G 08S 02 017 Socks, girl, size 8, short, style 2, material 017

Note the mnemonic aids, which make an awkward code but may be a great help in a busy, cluttered shop. Each facet of a code can be in the form of any code mentioned so far — sequence, block, or significant digit.

A decimal or hierarchical coding system uses principles similar to those of the group classification codes. A decimal point assists in identifying the major concepts. Such codes are capable of unlimited expansion by the addition of lower subsets. When a decimal code is used in a data-processing application, however, the maximum number of digits must be predetermined. If not, the design of files and subsequent retrieval from them becomes a very complex programming task.

The most obvious example of a hierarchical code is the Dewey decimal system widely used for classifying books. This code divides all knowledge into detailed categories, as it subdivides the following way—

Code	Items
3	Social sciences
37	Education
372	Elementary
372.2	Kindergarten
372.21	Methods
372.215	Songs and games
372.215.6	Action songs

The Dewey decimal system can be extended by the use of linkage symbols, such as a hyphen or an equals sign, to indicate degrees of relationship between separate code numbers. For example, heat treatment can be coded 621.785 and steel 669.14. A document coded 669.14-621.785 is concerned with the heat treatment of steel, because the hyphen linking the two parts signifies that the document deals with the first subject from the viewpoint of the second.

Alphabetic derived codes are of limited use in commercial applications but are sometimes found in information retrieval applications for document indexing and extensive name and address work. They are formed from the original alphabetic version of the title or name, abbreviated by the application of some standard set of rules. Examples of such rules are to remove all vowels from last names or to use the initial letter of the name followed by three digits formed from the second, third, and fourth consonants in it. For the latter, the consonants are divided into six groups:

Code Number	Letters Included
1	B, F, P, V
2	C, G, J, K, Q, S, X, Z
3	D, T
4	L
5	M, N
6	R

The letters *W* and *H* are ignored, and *Y* is treated as a vowel. This list may appear arbitrary, but it is based on phonetic principles. An advantage of these types of codes is that the reduction in the number of letters reduces spelling errors. The disadvantages are clumsiness and lack of uniqueness. Under the second rule stated above, for example, the surnames *Johnson* and *Jenson* are both allocated code number J525.

Although not directly concerned with coding, the address generation technique used to index random access files is relevant to the overall subject. (Section 5.3 describes the technique.) Indexing methods are covered in more detail in Appendix A, which deals with information retrieval.

A final consideration when designing a code is whether or not a check-digit system should be employed. A check digit is, usually, a single digit added to a code number to make it self-checking. It has a unique relation to the rest of the code number and its construction determines the type of error that can be detected by it.

For example, for the code number 315, a check digit may be formed by adding the individual digits of the number $(3+1+5=9)$. The check-digited code number becomes 3159. When writing or punching this number, if any digit is transcribed incorrectly (for example, as 4159), the number fails the test: $4+1+5=10$, not 9. Transposing the first two digits is another very common error. But if we write the number as 1359, the number still checks: $1\times3\times5=9$.

However, when designing a check-digit system, the type of errors that are to be detected must be considered. In the example above, if the first two digits are mistakenly transposed, the number still checks. In data processing, information is frequently punched into cards or paper tape from handwritten documents. Common types of error are:

1. Transcription, in which the wrong number is written, such as 1 instead of 7;
2. Transposition, in which the correct numbers are written but their positions are reversed, such as 2134 for 1234;
3. Double transposition, in which numbers are interchanged between columns, such as 21963 for 26913;

4. Random, which is a combination of two or more of the above, or any other error not listed.

Obviously, simply adding the numbers will not be a very good error-detection method. Most check-digit methods employ weights and a modulus. A weight is the multiplier used with each digit in the original code number to arrive at a product, and a modulus is the number used to divide the sum of the weighted products to arrive at a remainder. These terms are explained most easily by a simple example. Consider forming a check digit for a five-digit code number, using the weights 6-5-4-3-2 and modulus 11. If the original code number is 31602, the method of allocating the check digit is as follows.

1. Multiply each digit in turn by its corresponding weight. This gives (3 x 6) (1 x 5) (6 x 4) (0 x 3) (2 x 2)
          18   5   24   0   4
2. Add the resultant products $(18+5+24+0+4=51)$.
3. Divide the sum by the modulus and note the remainder $(51 \div 11 = 4$ with a remainder of 7).
4. Subtract the remainder from the modulus, and the result is the check digit. Thus the final step is $11 - 7 = 4$, and the new code number complete with the check digit is 316024.

To confirm that the new code number is correct, it should be multiplied by the chosen weight, using one as the weight for the check digit itself; add the results and divide by the modulus. There should be no remainder.

The modulus 11 method with weights from 10 to 2 is one of the most common methods used in data-processing systems. Its efficiency varies with the occurrence of random errors, but it will detect all other types of error. It is not the only method available, however; Table 3-F compares the efficiency of various ways of forming check digits. Table 3-G shows the results of a recent investigation within an installa-tion employing the modulus 11 method. Of 100,000 code entries examined, only 5 errors were undetected, resulting in 99.995 percent detection of errors.

TABLE 3-F
Efficiency of Check-Digit Methods

Modulus	Range of Weights That May Be Used	Max. Length of Number Without Repeating Weight	Weights Used	Percentage Errors Detected				
				Tran-scription	Single trans-position	Double trans-position	Other trans-position	Random
10	1-9	8	1-2-1-2-1	100	97.8	Nil	48.9	90.0
			1-3-1-3-1	100	88.9	Nil	44.5	90.0
			7-6-5-4-3-2	87.0	100	88.9	88.9	90.0
			9-8-7-4-3-2	94.4	100	88.9	74.1	90.0
			1-3-7-1-3-7	100	88.9	88.9	44.4	90.0
11	1-10	9	10-9-8....2	100	100	100	100	90.9
			1-2-4-8-16, etc.	100	100	100	100	90.9
13	1-12	11	Any	100	100	100	100	92.3
17	1-16	15	Any	100	100	100	100	94.1
19	1-18	17	Any	100	100	100	100	94.7
23	1-22	21	Any	100	100	100	100	95.6
27	1-26	25	Any	100	100	100	100	96.3
31	1-30	29	Any	100	100	100	100	96.8
37	1-36	35	Any	100	100	100	100	97.3

The analyst must be very precise in stating all details of the input when specifying his final system for the programming team. Each input for data preparation should be clearly

TABLE 3-G

SURVEY RESULTS OF A CHECK-DIGIT TEST

Number of Entries	Type of Error	Number of Errors	Errors Detected	Errors Un-detected	Percent Efficiency
EFFICIENCY OF MODULUS 11—ANALYSIS OF 1000 ERRORS					
100,000	Transcription	860	860	0	100
	Transposition	80	80	0	100
	Double transposition	10	10	0	100
	Random	50	45	5	91
	Total	1000	995	5	

named, followed by a description of the character code if more than one code has been specified in the general description. For input that needs translation into computer media, such as source documents to be punched into paper tape or cards, the specification should include a sample of the document (in an appendix) showing whether it is an existing document or a document specially designed for punching; the source of the document; a statement of whatever inspection, editing, transposition, or extension occurs before the document reaches the data-processing center; a statement of whatever presorting, batching, batch-numbering, and so forth occurs for control purposes; and a sample of any control slips produced.

The specification should also briefly describe the data preparation and verification system to be used, such as two-stage punch and verify, punch only, or read back and edit.

When the input is ready for entry to the computer, either directly or via data preparation, each input still should have an unambiguous name. It may be magnetic tape or disk files, magnetic ink character reader or optical reader documents, paper tape or punched cards from other programs or processes. For each type, the specification should state character by character, the format of the input, including the meaning of all acceptable control symbols; the content of every field, whether fixed or variable length, with or without leading zeros, and minimum and maximum possible values; the relationship between fields and control symbols; and the interrelationship of fields.

The specification also should insure that control symbols for temporary halts, such as that at the end of a reel of paper tape, are included. Compatibility of software labels should be noted.

Any system of input batching, control totals, and batch corrections must be described. Any method of correcting items within a batch, such as whether corrections are to be added to the end or shown subsequently, must also be described, and the system's data format shown.

The specification should also state where this type of input originates. For example, it may be output from an application or program already specified, or a by-product of an accounting machine invoicing run.

If, during its first input, data is to be checked by the computer, the specification should state —

1. Exactly what checks are to be performed for each input type, such as the formula of any validity check, the logical plan for any inter-field relationship check, and the rules for any checks on format, sequence, reference number, batches and batch control totals, or any other checks;
2. Exactly what action the computer will take on each error for each input type, such as outputting the faulty unit or batch of data; and
3. The manual procedure to be adopted for errors.

Different error procedures may be adopted for different input types. They can include omitting the faulty input from the current run for investigation and re-entry at a later date, or interrupting the computer while faults are investigated and corrected and re-entering input immediately. The latter procedure may be satisfactory if, for example, the computer has detected a sequence error during punched card input, or a damaged paper tape is being repaired.

If the computer is not to check input, the specification should state this. However, the analyst should remember

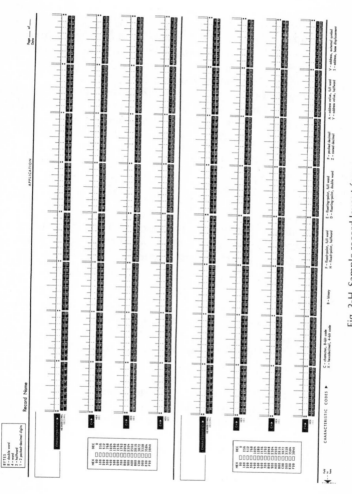

Fig. 3-H. Sample record layout form

that input of unchecked data results in output of question-able validity.

If punched cards are to be used for input, the specifica-tion should include a sample card design of the contents of each field. This is sometimes done on a blank card, but is often difficult with narrow column headings. An alternative is to use special card layout forms that show an oversize blank outline card and indicate the card columns by pre-printed numbers at equal spacing.

A record layout form such as the one shown in Figure 3-H may be used for the specification of tape and disk records. The layout record or supporting documentation should in-clude:

1. The application name
2. The record name
3. A list of the programs that use the file
4. The name of the program in which the file is created
5. A layout of the fields including file name, field mne-monic or abbreviation, field length, location of decimal points or other pertinent information, field character-istics, and maximum and minimum values for the field.

Exercises

1. Discuss the two basic categories of systems input.

2. Why isn't all data captured at its source in machine-processable form?

3. Discuss the advantages and disadvantages of any five types of input media. Why don't we simply standardize on a single input medium for all applications?

4. What factors are common to all types of input?

5. Contrast the advantages of punched cards versus paper tape.

6. What suggestions would you make about the design of a document from which cards were to be keypunched?

7. Briefly discuss the four main points at which input data can be validated.

8. What types of checks could be made on individual data fields as they are processed by a computer?

9. What are the objectives of a control system?

10. Explain why it is easier to control incoming documents in batches.

11. Explain three types of coding systems, and give an example of where each would be appropriate.

12. What is a check-digit system? When is it used?

13. What types of information should be specified when input data is to be checked by the computer?

14. Assume you are a systems analyst assigned the task of designing an order entry system for a wholesale drug company. You are trying to decide among the following methods of data capture the salesman could use as he visits each retail druggist to take orders:

   a) the salesman handwrites all information about the customer and the products ordered on an order form from which cards are then keypunched;

   b) the salesman mark-senses customer and product information on cards, which are then punched automatically;

   c) the salesman is given a deck of prepunched cards containing product description and price for each product; he simply marks customer information and quantity ordered on the face of the card, and this data is keypunched into the card;

   d) the salesman is given a deck of Port-a-punch cards and punches out all customer and product information as he takes the order.

Which methods would you choose? What factors would affect your choice?

output design

The output of the computer is the user's main contact with the system. Good output design helps to sell the advantages of the computer. A design that is unintelligible to all but its creator gives a poor impression of the entire system and may jeopardize relationships with users of the system.

The analyst must consider the various means of output available and choose the one that suits each particular application. In most commercial applications, the major output is provided by hard copy from a line printer, and printouts should be carefully designed to give the user maximum benefit from the system. Finally, the analyst must standardize the exact output requirements of the system so that both programmers and users know precisely what is to be produced. This chapter considers such details of output design.

4.1 Determination of systems outputs

There are three main subdivisions of output from a computer system:

1. External results, or the outputs for which the computer subsystem is to be implemented, are outputs that are to be distributed outside the user's organization, such as invoices, orders for suppliers, and accounts payable checks.

2. Internal results are the outputs necessary for the internal functioning of the user organization, such as control and audit listings, and error reports indicating where corrective action must be taken.

3. Internal system results are outputs such as operating statistics, run control and file usage reports, and system control reports.

The analyst must specify the following points for each output:

1. Identification
2. Content and format
3. Frequency and schedule of production
4. Volume of output expected
5. Conditions that trigger output production
6. Sequence in which output is to appear

Although a line printer is the most common output device, there are many others. The choice among them must be based on considerations similar to those used for determining input media. The points to consider are the response time demanded by the system, what quality of output is required, how many extra copies are needed, whether a hard-copy record is desirable, what peripheral devices exist or can be obtained, whether the system is integrated, and what the effective cost is. The following questions should be asked:

1. Is the cost justified by the benefit?
2. Are the user's requirements accurate?
3. Could two or more outputs be combined to provide a multi-use output?
4. Is the stated frequency necessary?
5. Is the response time shorter than necessary?
6. Should output be on demand rather than automatic?
7. Should the user have more control of contents and format and sequence?

The media available for providing output from the computer system are listed in section 12.1, which discusses input and output peripheral devices. However, the analyst should remember that some CPUs allow only specified peripheral units to be attached. For example, he must not design a system that requires a remote video display unit only to find that this particular device cannot be attached to his computer system. The manufacturer's technical data supplied with the computer specifies the range of output media available, which affects output design.

The analyst must consider how information is presented on output documents and what editing facilities will be needed in the programs. Output editing ranges from trivial tasks such as suppressing unwanted zeros and formatting lines correctly to such complex tasks as critically selecting records and combining them to report new figures. In the latter task, standard software packages called report generators can often help. At a still higher level, editing can become very complex when record selection has to be performed on variable or incomplete data. The job then becomes a problem of information retrieval, which is discussed in Appendix A.

The analyst should also use feasibility checks. For example, in a payroll run, a check that shows a negative net pay figure should require investigation. Finally, clerical output procedures should be considered from a standpoint of clerical efficiency.

4.2   Output record design

Most of the items discussed in chapter 3's section on form design are applicable to this section as well. However, continuous forms are more closely related to computer output than to input and will be discussed here.

It is wise to use the technical advice services offered by manufacturers who specialize in continuous forms. Feeding mechanisms on high-speed printers work within very fine limits of accuracy, and the forms must be printed with the same precision. Additional costs beyond the printing itself are involved in producing continuous forms because of the paper-handling equipment required (see also chapter 12).

It may be subsequently necessary to burst the forms (separate the continuous forms into individual pages). This is usually done at the same time as decollating (removing the carbons). The analyst should check that equipment is available to do this operation when planning the job and include the bursting costs in his estimate. Although multiple printing may save a small amount of time, the printout may not be usable until further time is spent in bursting. Even though bursting can be done offline, it may add several hours to the completion of a tightly scheduled job such as payroll. It is often helpful to mark the bursting position so that a constant check can be made. The size of the printout after bursting must be considered, since a narrow printout is difficult to handle without twisting or tearing and small documents are difficult to locate in standard-size files. Also, special folders for non-standard printouts are expensive.

If the total character width on the printer is a multiple of the maximum number of possible characters in the form, more than one report or several copies of the same report may be printed simultaneously on a single sheet. If a report is repeated twice across the paper it is called two-up printing. Two-up printing has many advantages but may lead to processing difficulties, since assembling several sets of data in storage and overflows from one form to the next may cause programming problems.

If the printout requires decollation and the decollator will simultaneously burst the forms, the saving of paper and time may justify two-up printing. If bursting must be done separately, single printing is generally best. Ease of programming and printer speed should be considered, and also whether printing is online or offline. Using two-up printing is especially valuable in installations that have a heavy printer load. Effective throughput speed, including printing, decollating, and bursting, should be the guiding factor.

The overall width of an output form should be kept to standard sizes to reduce paper costs and set-up time in form-feed devices, and to reduce filing and binding costs. Any vertical lines should split a print position. Otherwise, registration is difficult or impossible because of paper shrinkage and slight machine variations.

Line spacing on printers is generally either 6 or 8 lines per inch. Where greater readability is required, double spacing at 8 lines to the inch will yield 25 percent more lines per page than double spacing at 6 lines to the inch. It is often useful to make a sample study of the frequency distribution of the occurrence of body lines, such as items per invoice, before designing the form. It may be possible to save space and reduce the number of long skips or overflows to the next form. Skipping must be minimized, although it depends on the lines occupied by variable data. The total number of body lines should be a multiple of the spacing used. For example, if body lines are double spaced, the total number of lines should be evenly divisible by 2. Some lines require calculations before they can be printed and should be placed on the form so that printing is not delayed by calculation.

A preprinted form alignment guide mark, such as a small cross or dot, is helpful. If it is not present the operator will find some other way of determining when the form is set up properly, but this usually complicates set-up instructions and may lead to waste.

When marginal perforations are used, about one half inch should be allowed between the perforation and the first printed character. The practice of using form feeder holes for binding is increasing, particularly on internal documents. Usually, both sides of the form are untrimmed.

Preprinted form titles and other headings are neater and require less computer printing time; however, they are more expensive than blank stock and are used primarily as external forms and for the more important internal forms. Titles and headings for other internal forms may be computer-printed or supplied by a transparent template the reader overlays on each page.

The comments made about carbon copies in the previous chapter apply to continuous forms. If interleaved carbon is used, however, the output must usually be decollated before it can be issued.

4.3   Output standards

The analyst must precisely specify the output his system requires so that the programmer can provide what is needed. When output is in the form of magnetic tape, punched cards, or paper tape to be used for subsequent re-entry into the system, the specification should include the same details outlined in section 3.7 for prime input.

For printed output, normally from line printers, the specification should give each output an unambiguous name. If more than one code has been specified in the general description, it should state the character code to be used. For each printed output, the specification should include a sample of the output layout and tell whether it is on preprinted or plain forms. It does not have to include the preprinted design, but it should show areas where printing may take place and the location of each field. It should also allow for perforations, bursting, and other subsequent handling. An adequate margin should be allowed if the document is to be bound.

The specification should state whether each field is fixed or variable length; whether it is to include significant zeros, non-significant zeros, spaces, or other characters; whether it is left- or right-justified; whether it contains alphabetic data, or, if numeric, its maximum size and its maximum values.

If a form can have a variable number of similar lines, the permitted variation in number of lines should be specified and the procedure to follow if the maximum number is exceeded, as in the case of repeating page headings, should be noted. Line spacing should also be indicated.

If a form can be used in more than one way, sample forms and detailed statements of field layouts for all possible uses should be included. If line spacing before or after printing is possible, the analyst should specify which method is to be used.

For each output type, the specification should state the manual and offline machine procedures, such as whether documents are to be scanned to check print quality or checked in detail to prove accuracy of data, that are to be undertaken by the data-processing department.

What will happen to the output after it leaves the computer section should also be specified. It may be separated, folded, put into envelopes and delivered to the post office. It may be handed to the audit section or retained in the magnetic tape library for input to a subsequent job. The specification should also state who is responsible for accepting each output from the computer for subsequent use.

The analyst should include the print layouts as appendixes to the specification. These are usually drawn on line printer layout sheets supplied by the manufacturer. They contain a rule grid divided horizontally in one-tenth-inch areas and vertically with cross rulings one-sixth inch apart. Columns are headed by numerals ranging from 1 to the maximum number of print positions available. Since each square on the layout sheet corresponds to a print position, exact spacing is possible. (A minus sign occupies one print position.)

Another standard form that should be included in the specification is the output document specification sheet, which is illustrated in Figure 4-A. It is designed to support the more common print chart and link the output document with the storage locations and files that produce it. It is essentially an analyst-programmer communication, but, as part of the completed program documentation, is also a very useful programmer-programmer link.

The essential contents of the form are —

1. Heading

   Title of form
   Name of application to which it refers
   Name of analyst
   Date
   Form example reference
   Printer spacing chart reference
   Sequence of production of document
   Number of documents expected (average and maximum)
   Form size
   Printing size available

2. For each different type of print line to be output:

   Line number at which printing begins (as per layout form)
   Number of intermediate line spaces
   Final line number available for printing
   Average frequency (number of lines of this type per document)
   Source of output information
   Field/area reference in storage (for programmer's use)
   Field position (print columns to be occupied)
   Field name
   Fixed or variable length
   Maximum number of characters
   Editing requirements

OUTPUT DOCUMENT SPECIFICATION

Name_____

Sequence_____   Page_____

Form Size_____ Width_____ Depth_____ Number of Documents_____ Date_____

Print Area_____ Chars._____ Lines._____ Number of Documents._____ Prepared by_____

                                     Media ☐ Layout ☐

Line Numbers	Average Frequency	Source	Field/Area Reference	Field Position	Field Name	Fixed or Variable	No. Chars.	Editing Required

Fig. 4-A. Output documentation specification sheet

The system design may provide for log comments to be output on a console typewriter. Because of this device's slow speed, such output should be kept to a minimum and messages should be brief. It is sufficient to specify their content and meaning at relevant parts of the main text. For other types of output, particularly visual-display units, no standards have yet been developed because of their limited use. It is likely that a form similar to a print chart will be used for specification as these devices become more common.

Exercises

1. Explain the three main types of output from a computer system.

2. What points must the systems analyst specify for each type of output?

3. What considerations affect the choice of an output medium?

4. Discuss the advantages of laying out an output form on a print chart.

5. What disadvantages do forms with several carbon copies have?

6. When is two-up printing justified?

7. Explain the respective advantages of the two methods of printing form titles and headings.

8. Assume you are a systems analyst assigned the task of designing a sales analysis report for your company's sales manager. During your initial interview with him, what questions will you ask?

9. On a print chart, lay out a student grade report form, such as a student would receive after completing a semester. Specify which information would be pre-printed and which would be printed by the computer.

designing files

An office clerk usually needs access to more information than he can remember, so he keeps records of the information he needs. For example, a Kardex file is useful for a pricing clerk who is required to extend invoices every day. Customers' names and addresses, with details of their most recent purchases, are useful to a pricing clerk who also takes orders.

The computer suffers from the same problem as the clerk. The amount of information that it can hold in its working storage area is limited. Moreover, at different times in one day, the computer may perform the jobs of a pricing clerk, an order clerk, a stock control clerk, and a payroll clerk. So it adopts a solution similar to that of the office clerk; it holds the records for a given task in external storage.

The significance of files   5.1

Any ordered set of accessible records in a computer system is called a file. In particular, the term is used to refer to a set of records that must be retained over a number of operational cycles.

In a computer system, a distinction can usually be made between reference files and dynamic files. An example of a reference file is a file of product prices used to extend quantity by price to arrive at invoice value. The file is accessed to obtain information for the major task of the computer system. Changes such as price variations can be made in the file, but they are incidental to the major process.

An example of a dynamic file is a ledger file to which invoices are posted. Its major purpose is to record the constantly changing indebtedness of customers by entering each transaction as it occurs.

The way in which reference is made to the file or in which transactions arise to change the file helps to determine the storage medium to be used, the organization of records within the file, and the items within each record.

If records in a file are to be accessed at random, some form of direct-access device may be used. However, if the response time of the system is not critical, it may be preferable to sort the transactions into the sequence in which the file is organized. They can then be matched sequentially against corresponding records on the file.

The decision to organize the information in a given way on a particular storage medium is based on the following considerations:

1. The effect on the overall running time of the computer system
2. The availability or cost of the required storage media
3. The complexity of system design involved
4. The complexity of programming involved
5. The difficulties of operation

In most applications, file design is not the user's concern. It is usually resolved on the basis of the above technical considerations. However, the systems analyst must understand the file design since he may be responsible for advice on which policy decisions are based.

Both paper and magnetic media are used to hold computer files. Punched cards or punched paper tape are paper media; magnetic media are magnetic tape, disk, drum, and card. Of these, paper tape is used least. Punched cards and magnetic tape have been employed for a number of years, and magnetic tape has been preferred. Magnetic disk, drum, and card are very common now.

Choosing the best medium for the storage of a file in a given computer system depends on three characteristics.

First, the speed that information can be moved from the medium into the working area of the computer must be considered. Extremely fast internal processing speeds can be wasted if the time required to retrieve a record from the file depends on the relatively slow reading speed of some peripheral device.

Second, the possible ways in which individual records can be arranged for access on the chosen medium are important. Some applications require rapid access to particular records. In other applications, reading sequentially through all records on a file is suitable.

Finally, the volume of information that the medium can conveniently hold must be taken into account. Since a file may have to be extended to handle increased workload or changed system requirements, the medium's ability to handle expansion may be critical.

In every file design situation, there is a common problem: reconciling the structure of the information to be recorded with the inherent structure of the file medium. The punched-card file illustrates this problem very well. An individual punched card can hold a limited number of characters (usually 80). If no record in the file exceeds this length, one card will hold one record, and no difficulty arises. If an individual record contains more than 80 characters, however, more than one card is needed. Similarly, if record lengths vary, different numbers of cards are needed to hold different records.

A related design problem arises in determining how to arrange the component parts of the record. Consider the punched card again. If the required number of card columns is variable, there are two possible arrangements. In one, fixed-length fields can be allocated, but all but the largest items will waste space. In the other, variable-length fields can be used by adding special indicators to separate adjacent fields within the record. Using fixed-length fields facilitates programming, but it leads to inefficient use of the file, because even if a card column is blank it must still be read, and if the card reader is a serial one the time taken to read blank columns is wasted.

If variable-length fields are used, a high degree of efficiency in using the file may be achieved. However, related programming is difficult. The time saved in reading a compact file may be lost in breaking apart the fields in each individual record for processing.

These problems confront the systems analyst when designing any file in any media. He must carefully evaluate each systems requirement against the programming effort needed to achieve it.

Once the file has been designed, facilities must be available to add, delete, or change records so that the latest information is always shown. Such changes are called file maintenance. The procedures needed both to originate and to execute maintenance require the analyst's close consideration. Some techniques used for maintenance of three different file media are given in Table 5-A.

In some systems, it may be necessary to interrogate files. In interrogations, no content change occurs; the interrogator simply desires a report of current information as carried in a particular record or in a specific field in a record. File interrogation can be achieved most easily if a direct-access device is used. Very often, a console typewriter is used to input a message specifying the information required to the central processing units. The central processing unit determines the position of the pertinent record in the file, accesses it, and reads it into internal storage. From there, the required contents of the record are printed or displayed. If printing is used, either the console typewriter or a high-speed printer can be used.

With any file, regardless of the file medium, some method of locating records is needed. For this purpose, each record in the file must contain a key that identifies it. For example, the key in a sales ledger file might be the customer's account number.

The relationship of key fields in consecutive records is called the file organization. There are two methods of file organization: sequential and random. In a sequentially organized file, keys of consecutive records are arranged in a de-

TABLE 5-A

File Maintenance Techniques

Medium	Additions	Deletions	Changes	Method
Punched card	Punch new card; add to file in correct sequence.	Remove and destroy unwanted cards.	insert card with new data; delete card with old data.	Manual or offline
Magnetic tape or magnetic disk (sequential)	Copy existing records until key of new record reached; write new record; copy balance of file.	Copy records from old file to new file; omit unwanted records from new file, or insert a deletion marker on record.	Read record from old file, apply change, and output changed version on new file.	Computer
Magnetic disk (direct access)	Add new records to end of file, and change the location index.	Eliminate unwanted record with spaces, or insert a deletion marker.	Locate old record and write new version in same address.	Computer

fined sequence, usually from the lowest to the highest. A
random file has no defined sequence of key fields of consecutive records. The key of the first physical file could be
any key within the complete range.

Because of their nature, punched cards and magnetic tape
tend to be used for sequential files. The remaining three
primary magnetic media, although often used for sequential
files, are also capable of holding random files.

5.2   Punched-card and magnetic-tape files

Punched cards are not a very desirable file medium in terms
of speed, access, and volume. They have an effective transfer
speed of between 100 and 2,000 characters per second, which
is less than 1/5000 as fast as the internal processing speed
of the computer. Extracting selected records is usually impossible, and all cards in the file must be read for processing. A
large punched-card file is unwieldy and poses considerable
problems of clerical maintenance. Cards are bulky and occupy valuable storage space. For these reasons, punched
cards are normally used only for input and output, or for the
input and storage of programs. Their capabilities are too limited to permit extensive use as a file medium.

One result of punched-card limitations was the early introduction of magnetic tape as a file medium. This tape consists of a magnetic coating on a plastic base. The surface is
conceptually divided into very small grains that can be mag-
netized in one of two possible directions. One direction is
given the significance of zero; the opposite direction is given
the significance of one. It is thus possible to represent binary
digits that the computer can recognize and interpret by a
magnetic configuration in adjacent rows on the tape.

The amount of information recorded on the surface areas
of a tape is called its packing density. As many as 1,600 characters per inch can be recorded on magnetic tape, and a
complete reel can have a theoretical capacity of up to 35
million characters of information.

There is a discrepancy between the theoretical maximum
number of characters capable of being recorded on a tape
reel and the number it usually holds. This discrepancy arises
from the need to block information. The limited storage

capacity in the central processing unit requires data to be read in one block at a time. Each block usually contains several records.

The magnetic tape unwinds from one reel, passes a read/ write mechanism, and is wound on to a second reel. Once a block of information has been read, the tape is stopped to allow processing to occur. Before it halts, a small part of the tape travels past the read/write head. Similarly, when the next block of information is required, another small fraction of the tape passes the read/write mechanism before the correct reading speed is achieved. Areas known as interblock gaps, free of recorded information, must be allowed for stopping and starting. Consequently, the theoretical maximum packing density is never achieved.

Tape speed may be as fast as about 10 feet per second. Between one-half and three-quarters of an inch is crossed during the stop-start operation. With a packing density of 800 characters per inch, this amounts to space for 400 to 600 characters. Clearly, the larger the block of information, the more efficient the use of tape.

With a density of 800 characters per inch, and a tape speed of 150 inches per second, a theoretical data transfer rate of 120,000 characters per second is obtained. If the interblock gap is 500 characters and the block size is also 500 characters, the effective data transfer rate is about one-half the theoretical figure. The general formula for calculating the effective data transfer rate is

$$\frac{\text{Block size} \times \text{theoretical speed}}{\text{Block size} + \text{gap size}}$$

Table 5-B shows effective speeds for various block sizes, assuming a gap size of 500 characters.

Unfortunately, the bigger the block, the bigger the storage area needed to contain it. A typical block size is between 1,000 and 2,000 characters.

If the record size is less than the block size, it is usually desirable to block the records for the highest possible effective transfer rate. It is possible to adopt record layouts that involve variable-length fields. This in turn leads to the pos-

TABLE 5-B

EFFECTIVE MAGNETIC TAPE DATA-TRANSFER SPEEDS

THEORETICAL DATA TRANSFER RATE (CHARACTERS PER SECOND)	EFFECTIVE DATA TRANSFER RATE					
	With the Following Number of Characters per Block[a]					
	100	200	500	1,000	2,000	5,000
10,000	1,700	3,100	5,000	6,700	8,000	9,100
20,000	3,400	6,200	10,000	13,200	16,000	18,200
50,000	8,500	15,500	25,000	33,100	40,000	45,500
100,000	17,000	31,000	50,000	66,200	80,000	91,000
200,000	34,000	62,000	100,000	132,400	160,000	182,000

[a]Gap size assumed to be 500 characters

sibility of variable record lengths, and even of variable block lengths. Although these possibilities exist, one must consider the problems discussed previously so that programming is not inefficiently complex.

To appreciate this concept, one must understand the hierarchy of information on a computer file. The lowest level is an individual character of information, which is simply a given number of characters that the computer recognizes as a single logical unit. In other systems, the term *byte* is used—in which case, pairs of numeric characters, or one numeric character and its associated sign, or one alphabetic character or symbol can be packed into one byte of storage.

Because records stored on magnetic tape can be of variable length, and their constituent fields and resulting blocks can also be variable, special characters appear on the file to separate the various components. End-of-record and end-of-block markers are commonly used.

The following example shows an employee record that might be found on a magnetic-tape file:

1. Field 1 (1 character)—record type: employee name record
2. Field 2 (4 characters)—employee number, including department number
3. Field 3 (1 character)—sex
4. Field 4 (20 characters)—last name and initials
5. Field 5 (2 characters)—salary grade
6. Field 6 (2 characters)—year of employment

Each record would contain 30 characters, and ten such records might constitute a block. The use of the record type is required if the file also contains other employee records. In this case, salary details might be on the same file but in separate records because they are used by different programs.

A magnetic-tape file is normally processed as follows. Two main files—one called the old master file, and the other the new master file—are used. Transaction records are validated by a separate edit program that insures that each transaction meets the requirements of all subsequent processing operations. For example, a transaction record purporting to be shipping information might be rejected in an invoicing edit because no quantities had been entered for the items sent to the customer. Such rejections would normally be printed out on an error report, quoting details of the invalid data.

Valid transaction records are output by the edit program to form a transaction file on another magnetic tape or a direct-access device. It is then sorted to match the record sequence on the old master file. Transactions are matched one-to-one, and paired records are read into working storage. The transaction records are used to update relevant fields of the master records and the updated version is output to the new master file. At this stage, more errors could occur. For example, transaction data might be input for a nonexistent master-file record. Such errors would be reported on a file maintenance report. If no transaction record is present, unmatched records are copied from the old master file to the new master file unchanged. The new master file becomes the old master file in the next processing run. Figure 5-C illustrates the general flow of this typical file processing operation.

Note in this example that after it is initiated, incoming data must be sorted into the same sequence as the records on the master file after it is validated. If this were not done, the complete master file would have to be read to process each individual transaction, rewound to the beginning for the next transaction, and so on. Such a procedure would waste processing time.

One of the disadvantages of magnetic tape as a storage medium is its inability to update a particular record on a file at random. For example, if the required record is the hun-

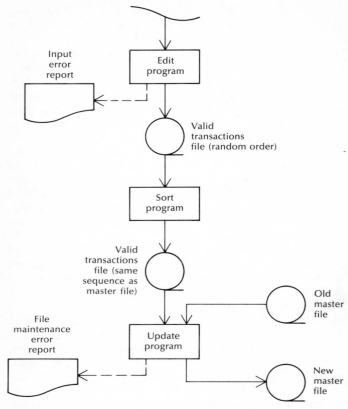

Fig. 5-C. Typical magnetic tape updating procedure

dredth in a file of 2,000 items, 99 records need to be read and copied on a second tape before the required record is reached. Then all the remaining records also have to be copied on the second tape after the updated item has been output. If the same record has to be updated again as a result of later processing, the operation of reading and writing 1,999 unchanged records must be repeated.

5.3   Magnetic drum, card, and disk files

Magnetic drums are expensive, but they have the fastest transfer speed of all direct-access devices. Their main limitation is a relatively small data capacity. For this reason they have been used primarily as an extension of the CPU's main storage. A typical use of a magnetic drum is for storing a segment of a large program while other segments of the program are in use, thus making more main storage available for working data.

Magnetic cards have a slower access time than disks or drums, but this disadvantage is offset by their very large capacity. They are much cheaper per character stored than either disks or drums.

Magnetic disks are in common use, and are second only to magnetic tape as a file storage medium. There are two types—fixed and removable—and fixed disks have a larger recording capacity. Removable disk packs consist of a number of circular recording surfaces, rather like phonograph records in appearance, mounted on a central shaft, as shown in Figure 5-D.

A disk surface in one typical unit contains 200 concentric recording tracks. The read/write head for that surface is moved in or out laterally to position on any required track. Once positioned, the head is stationary. Reading or writing takes place as the disk revolves. A complete disk pack consists of 6 disks mounted on a central shaft, and each disk except the top and bottom has 2 recording surfaces. The outer surfaces of the top and bottom disks are not used.

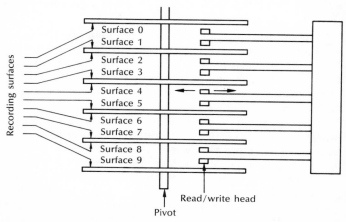

Fig. 5-D. Removable disk pack

The disk pack therefore has a total of 10 recording surfaces, each with 200 tracks. When in use, a disk pack is loaded into a unit providing a separate read/write head for each recording surface. A drive mechanism keeps the disks revolving continuously, and they are easily interchanged by an operator when necessary.

The recording surfaces within a disk pack are numbered from 0 (upper) to 9 (lower). On each surface, the tracks are numbered from 0 (outer) to 199 (inner). Each track is subdivided into a variable number of sectors which are the smallest addressable locations. Each track can hold up to 3,625 bytes of data, giving a capacity of 725,000 per surface, and over 7 million bytes per disk pack.

Data may be read or written at a peak rate of 312,000 bytes per second. The average time taken to position the heads and access any position on a disk is 75 milliseconds (approximately one-twelfth of a second). Since the access time quoted is an average, the actual time may be more or less according to the amount of head movement required.

In operation, all the read/write heads move in unison. Therefore, although data in the disk pack may be accessed in any sequence, the most efficient method is to access corresponding tracks on each surface in turn. For example, Surface 0 Track 9, Surface 1 Track 9,...Surface 9 Track 9, may be accessed in sequence without any movement of the heads following the initial alignment. Such a set of corresponding tracks is known as a cylinder (see Figure 5-E).

To minimize head movement, it is desirable to access all the records on one cylinder before moving to a new one. Allocating one quarter of a number of cylinders to one file, and the remainder of the *same* cylinders to a second file offers great advantages when the two files must be used in parallel during processing.

A disk file may be organized sequentially. In this case, it is processed like magnetic tape. Additions and deletions can be made only during a complete rewrite of the file. True direct access is not possible because there is no way of locating a record immediately on demand. Unlike magnetic tape, however, it is possible to find a record on a sequential disk file without reading each preceding record. This is achieved by a technique commonly called a binary search.

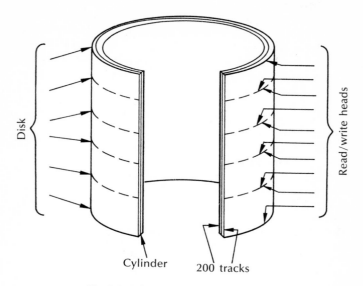

Fig. 5-E. Cylinder concept on disk files

With this technique, the middle record of the file is located first to determine if its key is above or below the required key. This automatically limits the search to half the file. The middle record of the selected half is then examined. Another check is made to see if the required key is above or below. By continuing this process of halving the number of records to be examined, the exact record is eventually located.

Although sequential organization does not use many of the advantages of a direct-access device, it provides the best solutions to many problems. It is typically used for files of tables or subroutines. Also, stored data is often arranged sequentially when a direct-access device is used for intermediate storage.

A modified form of sequential organization that is widely used on disk files is called indexed sequential organization. It offers both the advantages of a sequential file and the facility of random access when it is necessary to pass over a number of inactive records.

As the name suggests, the records are written in ascending key sequence on the tracks of the data area. Each record must have a unique fixed-length key. In addition to the data area, a cylinder/track index is also written — preferably beginning on track 0 of the first cylinder of the file. This index contains the highest record key in each cylinder and track.

When locating a particular record, a "high or equal" comparison of the record key against those in the cylinder index reveals the cylinder containing the record. The track containing the record is identified in the same way, and it is then searched sequentially to locate the record.

When processing an indexed sequential file, the overlay technique of updating is often used, which causes records to overflow within a track. To allow for this situation, an overflow area must be set aside. To minimize search time, this area should be on the same cylinder(s) as the data area.

It is sometimes necessary to locate items in the overflow area. To permit this, the cylinder/track index also records the highest record key for items displaced into the overflow area of each indexed track. This overflow address also indicates whether new file records should be introduced into the data area or the overflow area.

Record sequence in an overflow area can be maintained by a process called chaining. This method records the new location of a record in the overflow area within the record that logically precedes it. Similarly, the overflow record itself carries the location of the next record in logical sequence. In this way, the sequential order of records, wherever their particular location is on the disk area, is easily traced.

Indexed sequential organization has two main advantages over sequential organization for sequential processing. If a number of inactive records have to be passed, the next one required can be accessed directly without handling each inactive record. With a slight modification to the indexing technique, additions can be held in an overflow area assigned to each cylinder, thus avoiding the need to copy the file each time it is updated. Periodic reorganization runs can then restore the sequential order and copy the file for reconstruction purposes.

Since this method is common, the software provided by the manufacturers must have efficient indexing and overflow facilities. Many of the smaller disk systems allow indexed sequential methods to be used only if the file contains fixed-length blocks or fixed-length records. When variable-length records are part of the file, it is often necessary to rearrange the information by subdivision. An alternative is to fill records with spaces or blanks to make them the same length as maximum-length records.

Another type of arrangement is called a partitioned file. This method maintains an alphabetic directory as the index, and areas contain the records referred to in the directory. This type of file can be used to hold subroutines, for example.

Random file organization stores individual records in random order. Since no general rules for retrieving information from this type of file can be derived, no standard software is provided for it. The file user must write his own retrieval routine, which generally uses a full index or some address-generation scheme. Although random organization seems desirable, retrieval is often a slow process.

There are four basic methods of indexing—full indexing, partial indexing, self-indexing, and address generation. In describing differences among these four methods, the terms *bucket, block,* or *track* may be used to mean the area of the file that can be addressed individually. The analyst must apply the term used in his installation to avoid confusion.

Full indexing describes the situation in which an index record contains the bucket reference for every key in the file. The file is first sorted, and as it is transferred to direct-access storage, the necessary index is built. This index is sequential, although records may be stored in random order.

This method usually wastes space. The index normally gets too large to be held in internal storage. It is often written to another direct-access device and read into main storage in segments. Several accesses to the index may be necessary before the required record is located, and this can be a time-consuming process.

Full indexing is appropriate, however, where a high-activity portion of the file can be incorporated in the index itself. For example, the master file might be concerned with inventory. Each record could contain details of warehouse location, sources of supply, purchase prices, reorder levels, and so on. The information usually required could be the current inventory balance. In these circumstances, it might be practical to store the balance with the index. Reference to the master record would be unnecessary when requesting only this single item of information.

Partial indexing describes the situation in which two or more levels of index are carried on the file. The most common system is two-level: a rough and a fine index are maintained. The rough index contains the key of the last record in a given range with the bucket location of the fine index associated with that range.

With magnetic disk devices, the rough index normally contains a cylinder number. The fine index refers to a particular sector on a specific surface within the cylinder. Similar referencing principles apply to magnetic drum or card devices. The rough index selects a broad area that contains the given record, and the fine index pinpoints the record within the stated general area. This technique requires that individual records be stored in ascending key sequence within each bucket. The required record's key is compared against the rough index (on a magnetic disk file) to determine the cylinder that contains it. This cylinder is then accessed and contains the relevant fine index as its first record. It determines exactly where the record is within the cylinder, which is available without further read/write head movement.

At the beginning of each job, the rough index is usually moved into the internal storage area of the CPU and remains there while the file is active. In this way, individual file records can be accessed at very high speeds. The size of the rough index depends upon the file size and the number of natural ranges in the key fields of the data items. A limiting factor is likely to be the amount of internal storage that can be allocated for it.

Self-indexing, which consists of using the key of the record as the bucket address, is rare. Its use is limited because it often creates unacceptable addresses. However, if a file consists of fixed-length records in numeric order, self-indexing may be practical. For example, it might be used in a file in which customer account numbers are also the bucket addresses of the magnetic disk file of names and addresses. Thus, the name and address of customer 305785 would be found on cylinder 3, track 57, sector 8, in the fifth record. However, if there are significant gaps in the record sequence, self-indexing should not be used to avoid wasting valuable direct-access storage.

The fourth basic indexing method is address generation, which works on the principle of applying a mathematical formula to the key of the required record to identify the bucket containing it. The bucket is then searched to locate the record. In theory, this method should require only a single access for any given record. In practice, the formula is often out of date almost as soon as it has been introduced. It also often generates the same address for widely different keys. The bucket quickly fills, causing overflow problems. As a result, more than one access is often required to find a record. Another disadvantage of the method is that the formula rarely reallocates empty areas on the file caused by the removal of records, and space is wasted.

To produce a satisfactory address-generation formula, the pattern of the code numbers in present use and likely to be used in the future should be examined. From this analysis, a tentative formula that gives an even distribution of bucket addresses from the whole range of keys can be derived. Constants can be added in the formula where necessary to close large gaps in the sequence of the original keys. The resulting bucket address must be in a form acceptable to the particular direct-access device's hardware. The actual address-generation routine can be simulated on the computer and the distribution of the results analyzed to determine whether the formula should be modified. Eventually, a suitable formula for the program can be obtained.

Obviously, every formula must be individually designed to suit a system. Manufacturers' and other users' experience often indicates the best approach for a given series of keys. It should be remembered that to add, subtract, and shift numbers takes less time than to multiply or divide. Also, if alphabetic characters occur in original keys, the formula should cause them to be removed or replaced by numerals.

A randomized approach sometimes offers advantages. Randomizing means that successive records are not necessarily stored together on a master file. A sudden influx of sequential records is spread over the whole file and does not cause an acute overflow problem in one area.

A direct-access file can place records in either the main body of the file (in prime data areas) or in overflow areas. When this situation exists, processing time is used unnecessarily in searching two areas to find a given record. Therefore, at suitable intervals depending on the volume of changes, a file reorganization routine should be performed.

During this process, records are read and selected in sequence from the file. The sequential records are copied to an intermediate magnetic tape or disk which is then read back to the original direct-access device, thus reconstructing the file in sequential order. Since the reorganization outdates the old index, the software must index the new file.

In many situations, files are copied periodically for security. This function can be combined with file reorganization, and the intermediate file that is produced can become the required security copy.

Direct-access file size calculations are affected by the organization structure. The number of records must be multiplied by individual record length as usual, but additional allowances must be made. For example, records are sometimes arranged in fixed-length blocks. In other cases, variable-length blocks are arranged within fixed-length tracks. In either condition, maximum packing density probably will not be achieved and wastage should be anticipated.

Allowances may also have to be made for index areas on the file and control records for correct software interpretation. Overflow areas increase the total file length. It is usual practice to calculate a growth area for a disk file by setting a reasonable maximum number of records that may be stored. In this way, more than one file can be accommodated on the same disk and the starting point for successive files can be determined. With magnetic tapes, the problem of maximum usage is not so acute. The relatively low cost of tapes allows them to be used flexibly. A small unused area on a magnetic-tape file may be economic from processing considerations alone.

Copying and overlay are two methods of updating a direct-access file. Updating by copying is the technique used for magnetic-tape files. The old version of the file is read, necessary processing is performed, and a new version is created on a separate storage area. The old version remains untouched by this process. Updating by overlay is unique to direct-access files. The new version of the file is created by changing the records on the original file, and the old version is destroyed.

There is a use for each method, depending on the application. For example, if each record has to be updated in a run, it is probably better to use the copying method. If only one percent of the file is accessed in each run, updating by overlay is probably preferable.

Sequential or random processing can be used to process a direct-access file organized on a sequential or a random basis. Sequential processing means arranging input data in key sequence. Random processing means accepting input data in any sequence. Table 5-F describes the four methods possible.

Since there is a choice of processing methods, the analyst must decide which system is best for the application under review. With sequential processing, input data must be sorted before actual matching with the files can begin. However, once it has been sorted, minimum read/write head movement is required to perform the actual processing.

With random processing, although input data requires a minimum of sorting, considerably more accesses to the file are necessary. The required records can appear anywhere. In some jobs, additional read/write head movements for locating records may make the overall processing time longer than if presorting and sequential processing were performed.

TABLE 5-F

COMPARISON OF DIRECT-ACCESS PROCESSING METHODS

FILE ORGANIZATION	PROCESSING METHOD	
	Sequential	Random
Sequential	Often the fastest when batches of transactions have to be processed. Is the method used with magnetic-tape files.	The method often used for real-time systems. Can be achieved by indexed sequential organization. Suitable when small batches must be processed, or several files must be updated by each transaction.
Random	Useful when variable-length records have forced random organization. Each transaction record must be put through address location routine. Results are sorted on generated address rather than key.	A rare technique, but can be used sometimes for table look-up procedures.

Another technique sometimes employed when direct-access devices are available is multiple or "one-shot" processing. Input data is read only once and used to update more than one file before any new input data is considered. A system may sometimes require this approach when the effect of one transaction must be known to give the correct result on the next transaction. Some production control routines are in this category. The technique may offer advantages over sequential processing by avoiding the need to batch and sort input data into different sequences for different files.

5.4 Sorting and merging

Sorting is a basic technique common to all forms of data processing; it implies the rearrangement of information into a predetermined sequence. Merging is combining of two or more sorted lists into one list. By implication, it is assumed that the resulting merged list will also be in a given sequence. Since up to 40 percent of a system's overall processing time can be spent in sorting and merging, it is important to perform these operations efficiently.

Sorting depends on the ability to make comparisons. During a sort, two objects are compared and placed in sequence by reference to some standard. For example, when sorting single figures into ascending sequence, the normal decimal conventions apply. 1 comes before 2; 2 comes before 3; and so on. When unfamiliar symbols such as the signs of the Zodiac are used, sorting by mechanical methods is impossible because a standard giving the order is unknown. Thus machine sorting is possible if the machine is capable of recognizing and comparing patterns in a file. These recognizable patterns are the key fields in records.

The analyst must be alert to ways of reducing machine sorting time. One method is to introduce preliminary sorting of input documents before transcription into computer input. Color, code letters, layout, and so forth are aids to preliminary offline sorting by clerical routines. If the system is carefully designed to accept sequenced or partially sequenced input, overall processing time will often be reduced by faster internal sorting.

Because of its complexity, mechanical sorting theory has been subjected to mathematical analysis. The following discussion is a simplified introduction to the subject.

Radix, or distribution sorting, is the technique that has been used for many years in mechanical punched-card sorters. It allocates a suitable number of "pockets"—10 for numeric keys and 26 for alphabetic keys—and then distributes the incoming unsorted data into relevant pockets. An item's destination is selected by examining each character in its key, starting with its least significant character. Figure 5-G illustrates the technique.

ORIGINAL
LIST:
48
32
69
21
03
15
06
01
10

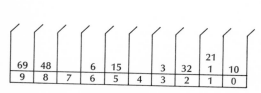

Pass 1: Distribute on units column.

RESULT OF
PASS 1:
69
48
06
15
03
32
21
01
10

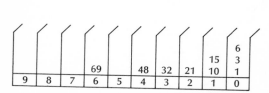

Pass 2: Distribute Pass 1 results on
tens column.

RESULT OF
PASS 2:
(sorted list)
69
48
32
21
15
10
06
03
01

Fig. 5-G. Distribution sorting

Sorting by insertion is another technique in which each key is examined in full as each record is read. The first key begins a list. Subsequent keys are inserted in their correct places, and existing members are pushed down if necessary. An example is shown in Figure 5-H.

ORIGINAL LIST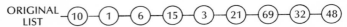

Each number is read in turn and inserted in the list. Numbers already present are shifted to the right to make room. Numbers in double circles are input numbers on each line. Arrows show movements.

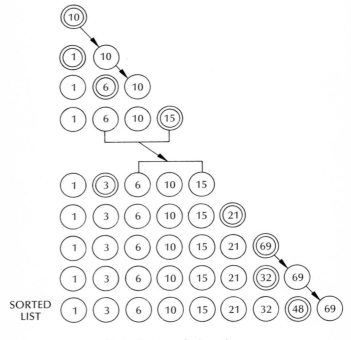

Fig. 5-H. Sorting by insertion

Address calculation is a method that can be used with small lists. It assumes that the range of keys is fairly well known. An estimate is made of the final location of each record in the sorted list by referring to its key. The number of locations needed to hold the results during the sorting operations should be in the ratio of 2·2:1 for minimum shifting. Thus, for a list of nine numbers, twenty locations are required.

Another method, which requires relatively little internal storage and reduces the number of comparisons, is known as selection sorting. The total number of items to be sorted is divided into subgroups. The number of subgroups and the number of records within each subgroup are the same. Each is equal to the square root of the original total number of items. For example, sixteen items are divided into four subgroups of four items each. By repeated extraction of the smallest key from each subgroup, as illustrated in Figure 5-I, a final sequential list is produced.

Sorting can also be performed by merging, if each member of any sequence is considered as a presorted list of one item. With a simple two-way merge, the original records are read from one tape in pairs, sorted into ascending order within the computer, and output on alternate work areas on some peripheral device. The records on the work areas are referred to as strings. They are read from the two areas, merged into groups of four, sequenced again within the four, and output to two other work areas. Each complete reading and output operation is called a pass. The passes continue, merging strings into longer strings each time and writing the results to the work areas alternately. Eventually, one merged list is created. Figure 5-J demonstrates the process. Multi-way merges are performed by employing more work areas. The number of passes rises rapidly, depending on the number of strings originally involved.

Sorting programs generally require two phases: forming strings of sequenced records on some external storage media and repeatedly merging them until a single sequence string is produced. The formation of the original strings of records is performed by the sort software and often uses the "replacement-selection" technique. Main storage is initially filled with multiples of four records in random order. Pairs of adjacent records are compared continuously until the record with the smallest key is detected. This smallest key record is then output. A new input record replaces it in the identical physical position. An advantage of this method is

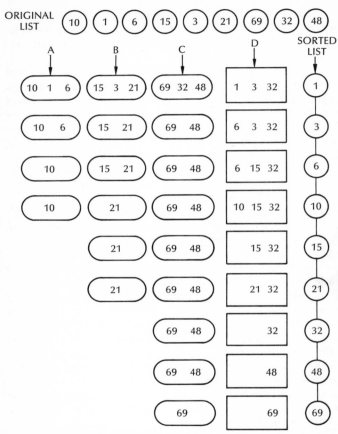

Method: Form 3 subgroups A, B, C. Select smallest number in A, B, and C to form group D. On each step, take the smallest number in D as the next member of the sorted list. Replace the extracted number from D with the next smallest number remaining in the subgroup to which the extracted number belonged.

Fig. 5-I. Selection sorting

that reading, writing, and comparisons can take place simultaneously if the computer permits overlap of these functions.

The distribution of strings on the external storage medium (usually magnetic tape) is dependent on the type of merging

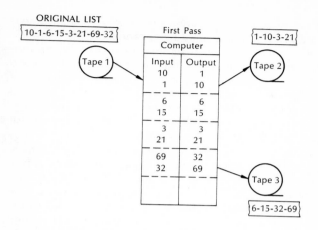

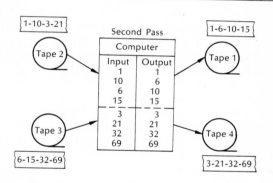

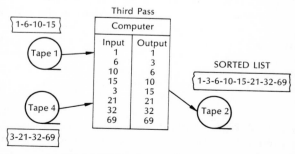

Fig. 5-J. Sorting by merging

to be performed as the second phase. In a standard n-way merge, the distribution of strings is equal. In the more advanced technique, known as a polyphase merge, the distribution of strings is unbalanced. This latter distribution is based on either the Fibonacci sequence of numbers or a sequence of numbers built in the same general way. The Fibonacci series is simply a set of numbers in which the current number is equal to the sum of the two preceding numbers, for example 1, 2, 3, 5, 8, 13, and so forth.

Once the strings have been distributed, records are merged until a complete single string of sequenced records is produced. The techniques and algorithms necessary to produce these results are the concern of the software designer. Analysts interested in the technical aspects are referred to the publications listed in Appendix C. The programmer is commonly required to know only the parameters to be input to the standard sort/merge software package supplied by the manufacturer. These are quoted in detail in the appropriate reference manual for his particular machine.

When sorting and merging on a computer, some problems arise. One of the most important is the question of key recognition. The actual part of the record that is to form the sort key must be clearly specified. For ease in sorting, the key should be in the same relative position on every record. This point requires close attention if variable-length records are used.

The computer must also be instructed if some special sorting order is required, which occurs whenever the information is not to be sorted in strict numerical or alphabetic sequence.

Finally, since internal storage is required for sorting processes, the amount of information that can be processed at one time is limited. For this reason, peripheral devices are used as temporary storage areas. The overall sort is segmented into a number of much smaller sorts, each performed within the internal storage area to form strings of sequenced records that are stored temporarily on a disk or tape and are finally merged to form one sequential series.

The main advantage of disks over magnetic tape for temporary storage is that they allow the user to access any required string after the first allocation has been made. This often allows strings to be increased by reading them from any required position on the disks; it can lead to faster sorting. On magnetic tape, rewinding and rereading the strings would be required. The advantage can be partially offset, however, by the problems encountered with the relatively slow arm movement on disks. Transfer time can be wasted while the arm is moving over the revolving disks to locate required strings.

Difficulty is sometimes experienced when records to be sorted are very large. One solution is to divorce the main record from its sort key and simply to tag the sort key with an address of the storage location of the main record. The keys and their associated tags can then be sorted into the required sequence. When sorting is complete, the sequenced keys can be output. They can be matched to their main records again by reference to the tag.

No general timing formula can be given for sorting, because read/write time depends on block size. In turn, the number of blocks that can be handled depends on the input and output areas available in internal storage. All manufacturers supply sort-timing tables for their machines. The analyst can use them to estimate the time required to sort a given number of records of a specified average number of characters in length.

Because of the inherent complexity of sorting methods, manufacturers provide standard sort routines as part of software. This software must be held in internal storage to control sorting and merging. Its size limits the space available in which to carry out the processes themselves.

In general, the parameters the user must specify include the position of the first key field, the overall length of the key field, whether records are fixed or variable length, whether the job is complete in itself or whether the final operation is the return of control to the user's program, and so on. Most sort programs allow the user to build actions into the sort operation as the system requires. Additional coding routines are normally called user routines. For example, a user routine might be used to delete unwanted records during a merge. As with timing data, the scope of sort routines varies among manufacturers.

5.5   File standards

Usually file data is stored magnetically, but it also may be stored as punched cards or paper tape. For each file referenced within a program, the detailed specification prepared by the analyst gives the file name, states the medium to be used for storing the file, whether all records are similar, or whether records of different types are to be grouped on the file. If the latter condition is true, provision should be made for blocks of control data throughout the file. The key system used to determine file sequence should also be indicated.

For subsequent program planning, the specification should state an approximate total volume for each file, in terms of numbers of records and total storage capacity required. The names of other files that may occupy other parts of the same reel of tape, disk pack, or other storage device should be given. Whether records are fixed or variable length should be noted.

For variable-length records, the analyst should specify whether the complete record must be accessible in main storage during processing. This affects the packing method to be used within magnetic-tape blocks of disk tracks. For a record made up of a header and trailers or subrecords, which are repeated groups of fields, the sequence and the frequency of trailers must be stated.

In addition, for each record type, the specification should state the name and data level of each field, and whether it is a reference number, a descriptive code symbol, a control count (such as number of words in the record), or normal data. It should show whether the field is fixed or variable length, and the fixed or maximum number of characters. It should also state whether the field is numeric (giving minimum and maximum algebraic values) or alphanumeric (stating the field length). The number of like fields within the record, either a fixed figure or maximum and average values for a variable figure, should be quoted, and the relation between different fields should be given.

During processing, certain combinations of input data, dates, file data, and program parameters will cause file information to be referenced or altered, and files or output to be created. Any rules of calculation or formulas to be used should be stated. All symbols to be used should be defined.

For each run, the specification lists conditions which can cause maintenance of each file, including such factors as "year-end" cleaning up. The specification, therefore, should state:

1. The frequence of maintenance, such as daily posting, weekly housekeeping, and annual revision;
2. The source of all data forming each new record to be inserted and its position within the file;
3. Whether a complete record is to be deleted from the updated file, and if so, whether such records are to be ignored, printed on an output listing, or written to a separate file;
4. Whether a field or group of fields is to be inserted, deleted, or changed;
5. Whether file maintenance is a combination of those above factors, and what all file modifications are to be;
6. The effect each type of maintenance has on control data; and
7. What priority system is to be used if more than one modification occurs during one process.

The computing technique to be used during file-maintenance runs should be stated. In any situation other than sequential organization and processing that uses separate old master and new master magnetic files, a full statement of the method and special programming considerations should be included. This statement is particularly relevant to random and real-time processing.

If father-son sequential updating is to be used, it should be specified. The frequence with which a file can be updated, the points at which backup copies are to be retained, and the number of generations to be retained should also be listed. If any other backup method is to be used, the file security system, which may appear in other parts of the specification, should be emphasized. A statement of file-recovery procedures, both for rewriting a complete file and for a faulty individual record, should be included.

The operating system to be used for a set of programs, whether it be entirely in line with software, or entirely "load and go," or at some point between the two extremes, must be specified. Furthermore, if a console typewriter is to be used for any purposes other than standard software uses, it should be stated. The messages are defined under the input and output sections.

If a specification is written for an application permitting independent programs to occupy separate parts of the same computer at the same time, a special note should be made. For such multiprogramming, the specification should state what hardware and software limitations are imposed, and what the hardware and software processes are, including dumping special registers and linking, branching out of a program, and returning to a program.

The file record specification, like the other standard specification documents for input/output, supports the written program specification. The essential contents of the form are:

1. Heading

    Title of form
    Name of application
    Name of analyst
    File reference name
    File sequence
    Date
    Layout reference
    Record length (maximum and average)
    Block length
    Number of records (maximum and average)

2. For each field

    Field area reference (usually mnemonic program name)
    Field position in record (modification factor from start of record)
    Field name
    Fixed or variable length
    Number of characters (maximum)
    Mode of characters (binary, alphabetic, numeric)
    Maximum values (including allowances for packing data, and so forth)

A sample document is shown in Figure 5-K.

Finally, the specification may include some reference to

FILE RECORD SPECIFICATION

Title _____ Application _____ Page _____

File Reference _____ File Sequence _____ Date _____

Record Length _____ Maximum _____ Number of Records _____

Block Length _____ Average _____ Prepared by _____

Layout ☐

Field/Area Reference	Field Position	Field Name/Description	Fixed or Variable Length	No. Chars.	Mode	Maximum Field Size	Maximum Frequency	Totals

Fig. 5-K. File record specification

main storage and auxiliary storage requirements. The analyst is concerned only if look-up tables that form an integral part of his system may have to be held in core, and he must leave space for the program and working data. These problems are normally resolved by analyst-programmer liaison. Solutions are noted in the specification for reference.

Exercises

1. Draw a flowchart showing the procedure to be followed to make an average of 500 code changes to the keys of records contained in a magnetic-tape master file. The changes may be up or down. Input is from paper tape.

2. List the fields you would expect to find in a record master file used in a personnel/payroll application. Quote the number of characters in each field and state whether they are of fixed or variable length. Arrange the fields in logical order for processing. Draw a file layout for one complete record on magnetic tape.

3. A file of records is to be held on a magnetic disk that has 3,625 bytes per track. Each record is 160 bytes long. The maximum size of the input area available for this file is 1,000 bytes. How many records per block would you choose, if the prime requirement is that records are packed as densely as possible?

4. A file area of 200 buckets has been allocated on a disk device. The file consists of fixed-length records that can be loaded at 24 records per bucket. The keys of existing records are 00516 to 02916, 04131 to 04530, and 10001 to 11800. Additional records will be in a new group starting with 04531. The user of the system requires that these new records follow 11800 in any printed reports. Reports printed by the computer are mainly in sequential order. Draw a flowchart of an addressing system that can be used as a subroutine for record location.

5. What is the difference between maintaining and interrogating a file?

6. Why is it more efficient to record data on magnetic tape in multiple-record blocks?

7. Contrast magnetic drum, card, and disk files in terms of speed of access and data capacity.

8. Why is it more efficient to organize a file vertically (from disk surface to disk surface), rather than horizontally (from track to track on the same disk surface)?

9. Explain any three sorting techniques. What considerations affect the choice of a given technique?

10. Assume you are a systems analyst assigned the task of selecting a file organization method for an inventory master file of 12,000 records of 95 characters each, stored on magnetic disk. Orders for stock come in randomly from high-speed terminals located throughout the assembly plant, and whether there is sufficient stock on hand to fill each order must be determined immediately. All reports prepared from this file are in stock number sequence. Which method will you choose, and why?

6

design and documentation

By studying a system's input, output, and files, the analyst discovers the processing and computational requirements that are necessary to provide the links between them, and thus creates a new system design. He then has the responsibility of explaining the design to management and securing their approval.

While the analyst explores alternative solutions to the problem at hand, he is largely a free agent. His own ingenuity controls his activity. However, after he resolves the solution and must communicate it to programmers and users, his freedom is restricted by his installation's documentation methods.

Though standards for documentation may vary widely from one installation to another, only one method can be employed within any one organization. This chapter examines some common problems of systems design and discusses factors that influence the establishment of a standard method of documentation.

Considerations in system design 6.1

When the analyst designs the actual processing and computa-
tion of data in his system, he faces numerous problems.
Many of these are due to limitations imposed by the user or
in the installation's hardware and software. Some of the
matters that will require his attention are discussed below.

1. A *one-input system* may be a small routine or a large set
   of programs. The stage at which a function becomes
   large enough to justify making it a separate procedure is
   usually determined by the amount of coding involved.
   Such systems may arise, however, when the function is
   particularly complex and/or requires specialist knowledge *or* when the appropriate routine or program already
   exists in usable form.

2. A *two-input system* commonly implies matching by a
   sequential method if both input streams are in the same
   sequence, or by a direct-access method. In the latter
   case, records to be matched against are available on a
   direct-access medium with short access time, and transactions are dealt with on arrival and not necessarily in a
   predetermined sequence. This method is required in a
   real-time system. A variation of the sequential method
   is to take several groups of records at one file, and for
   each such group of records, to pass the entire other file.
   Each group of records is held in main storage (or possibly on a direct-access medium). Each pass of the transaction file processes all transactions applying to the records held in the current group of records. It may be wise
   to rewrite the transaction file during each pass, dropping
   transactions that have been matched. This method can
   minimize sorting time. Even when it is possible to deal
   with one or both files in random sequence, this procedure may be erroneous or unacceptable because a
   definite order is required for several transactions applying to the same record (for example, one must change a
   record before using it to answer an interrogation). In
   this case, the transaction file must be sorted.

3. *Multiple-input.* Generally, it is inconvenient to handle more than three inputs in a single program. In particular cases, however, a multi-input matching procedure may be desirable. Usually, the number of peripheral devices determines the number of input streams that can be handled. If only two inputs are available, a system of merges or multi-input sorts must often be used. Careful consideration may be needed when deciding whether to execute a lengthy merge program immediately before the two-input run or to sort and merge the several input files progressively as they become available.

4. *Spooling and Media Conversion Programs.* "Spool" = Simultaneous Peripheral Operations Off-Line (tape-to-printer, card-to-tape, and so forth). These programs normally impose certain format or sequence limitations. However, these limitations are usually acceptable because of the advantages obtained from using proven programs and standardized operating procedures when handling slow peripheral devices.

5. *Multiprogramming.* The operating system may impose significant limitations on the way multiprogramming facilities are used. A "foreground-background" system with partitioned main storage may be the basis on which the machine has to work. A full 15-program system or a multi-access system allowing 200 remote users to work concurrently may be available. Within the manufacturer's limitations, the installation operating management may impose restrictions of their own, designed to insure that the machine's facilities are profitably used.

6. *Real-time.* Input, output, and process with access to files must be handled immediately. Each procedure is performed only once before going on to another; work is by transaction rather than by process (as in batch processing). Usually all types of input transaction are dealt with in one message-handling program that edits and examines the message. The message defines the procedures required and the files to be accessed. The message-handling program deposits the decoded message

data in a standard area and initiates a transfer of control to a processing program. This program or sequence of programs deals with the message and deposits the results in another standard area. An output program passes the results to the remote terminal. The processing program may have passed on details of the message for further processing. The manufacturer usually provides programs or sets of routines from which programs can be assembled to handle communications tasks. These programs impose certain standards of internal communications on customer programs. The communications package deals with context-independent checking, line failures, queuing problems—all of which are specialist topics.

7. *Run segmentation.* Segmentation of the total task is the crucial stage in system design. The basic task is to identify the division of labor between computer and non-computer. Generally, the smaller the number of programs the better. However, from an operational viewpoint this statement is a gross oversimplification. It is sometimes necessary to segment because programs would otherwise be too large to be accommodated in main storage, and a sorting function implies a split in procedures (for example, in Data Edit-Sort-Update).

8. *Optimization.* In section 1.4, it was stressed that the design process is iterative. After the analyst completes the initial design scheme, he must examine the proposed design for weaknesses. He must pay particular attention to questions of peripheral or processor domination and file design. The most straightforward methods of optimization are run combination and run segmentation. Run combination reduces setup time and may eliminate one or more passes of a file. If a computer-limited run can be combined with a tape-limited run, a balanced combined run may result. Segmentation of runs tends to increase setup time, but it can be used to reduce demand on peripheral devices. In some processor-bound runs, processing time may be reduced by partitioning into two balanced runs or by moving a particular processing function into the next program.

9. *File design* must be examined carefully to make sure the right degree of compactness has been achieved. If one or more runs handling a given file are file-limited, it may be wise to try to reduce the physical sizes of the file records, thus reducing file-passing time. Such compacting tends to increase processing time, since extra coding must be inserted in the affected programs to unpack the compacted fields. Conversely, if runs are processor-bound, it may be wise to expand records, thus easing the task of unpacking the record fields. If a system is to be run in a multiprogramming environment, it is highly desirable to avoid processor-bound runs that tend to limit the system's flexibility. Activity distribution on files should be considered from the beginning. Any "80-20" situation (where 20 percent of the master file records receive 80 percent of the transactions) should be detected, and the file design should be modified accordingly.

10. *File buffering* may be controlled by the programmer. Nominating one, two, three, or more buffer areas for a file may be useful, particularly where a list requires significant processing time but the activity of the file is very low. A large number of buffers can then be employed to maximize the amount of overlapped file time, by allowing a queue of blocks to build up while the list is being processed.

11. *Sorting runs* almost invariably use the standard sort program supplied by the manufacturer. The use of first and last pass "own-coding" options should always be considered. In an extreme case, this can save a whole run. If the manufacturer's sort program is fully understood, it may be possible for the preceding program to produce data so that the first pass of the sort is easier and faster. Similarly, if the first pass of the sort is processor-bound, a reduction in the storage this pass uses to form strings may balance it. In critical situations, a pass in the second phase may be eliminated by reducing record size and increasing string length out of the first pass.

Many of the limitations and problems mentioned above are deliberately programming-oriented. They have been included to show the analyst how necessary it is to work closely with the programming team during the design stage. If he ignores the programming implications of his design, he may find later that the equipment cannot handle it. Consulting with a senior programmer who can quickly point out what will cause programming difficulties will save time and effort.

Flowcharts 6.2

After the analyst has recorded all the relevant facts, he studies them and produces a tentative design for the new computer-based system. This first draft is discussed and modified. After several drafts have been discussed, the analyst specifies his system in a manner suitable for programming work to begin.

The most common form of representation of data-processing functions is the flowchart. The systems analyst uses flowcharts to plot the flow of information in and around his system. (Aspects of flowcharting have been discussed in chapter 2.) There are two typical levels in systems design flowcharting: system flowcharts and program flowcharts.

A *system flowchart* shows the flow of documents around the computer system and outlines the functions or processes performed by the computer. Computer input and output (for example, paper tape and punched cards) are not usually shown at this level.

A *program flowchart* is the most detailed flowchart drawn by the systems analyst. It supports the written program specification and is used to clarify processing procedures within a program. It corresponds to the highest level of flowcharts a programmer draws. The input, output, and file layout documentation referred to in previous chapters support this flowchart in the final program specification.

Figures 6-A and 6-B show a typical system drawn at both levels, and Figure 6-C illustrates symbols used in both kinds of flowcharts.

Observe the following principles in drawing flowcharts:

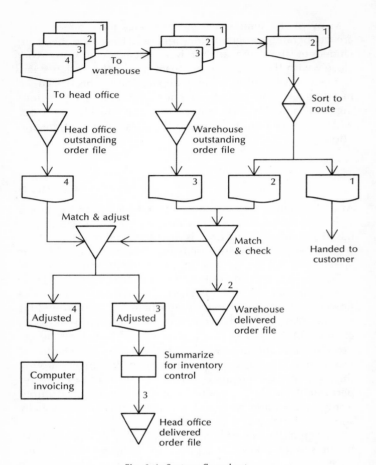

Fig. 6-A. System flowchart

1. Make the flowchart clear, neat, and easy to follow.
   It will then make a good visual impact and communicate well.
   (a)   Mark the logical start and end points.
   (b)   Use standard flowcharting symbols.
   (c)   Avoid crossed flow lines.
   (d)   Use simple decisions, that is, those giving yes/no
         or greater than/equal to/less than answers.

(e)   Work in a consistent direction down or across the
      page.

2. Be logically correct.
   (a)   Do not miss actions; account for all branches.

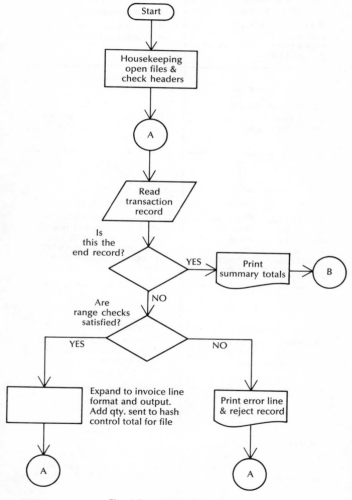

Fig. 6-B. Program flowchart

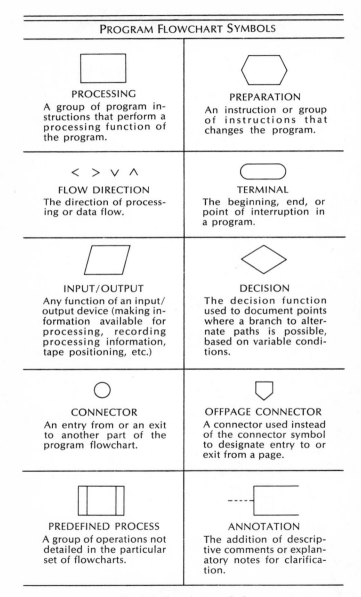

PROGRAM FLOWCHART SYMBOLS

PROCESSING
A group of program instructions that perform a processing function of the program.

PREPARATION
An instruction or group of instructions that changes the program.

< > ∨ ∧
FLOW DIRECTION
The direction of processing or data flow.

TERMINAL
The beginning, end, or point of interruption in a program.

INPUT/OUTPUT
Any function of an input/output device (making information available for processing, recording processing information, tape positioning, etc.)

DECISION
The decision function used to document points where a branch to alternate paths is possible, based on variable conditions.

CONNECTOR
An entry from or an exit to another part of the program flowchart.

OFFPAGE CONNECTOR
A connector used instead of the connector symbol to designate entry to or exit from a page.

PREDEFINED PROCESS
A group of operations not detailed in the particular set of flowcharts.

ANNOTATION
The addition of descriptive comments or explanatory notes for clarification.

Fig. 6-C. Flowchart symbols

SYSTEM FLOWCHART SYMBOLS

PROCESSING A major processing function.	INPUT/OUTPUT Any type of medium or data.	PUNCHED TAPE Paper or plastic, chad or chadless.	PUNCHED CARD All varieties of punched cards including stubs.	ONLINE STORAGE	MAGNETIC TAPE
DOCUMENT Paper documents and reports of all kinds.	DISPLAY Information displayed by plotters or video devices.	TRANSMITTAL TAPE A proof or adding machine tape, or similar batch-coded information.	AUXILIARY OPERATION A machine operation supplementing the main processing function.	MANUAL INPUT Information supplied to or by a computer utilizing an online device.	MERGE Combining two or more sets of items into one set.
KEYING OPERATION An operation that uses a key-driven device.	MANUAL OPERATION A manual offline operation not requiring mechanical aid.	OFFLINE STORAGE Offline storage of either paper, cards, magnetic or perforated tape.	COMMUNICATION LINK The automatic transmission of information from one location to another via communication lines.	COLLATE Forming two or more sets of items from two or more other sets.	SORT An operation on sorting or collating equipment.

Fig. 6-C. Flowchart symbols (cont'd.)

(b)   Do not repeat actions.

(c)   Show a complete logical flow from start to end.

3. Work at a consistent level of detail. If a particular action implies a great deal of processing at a lower level of detail, draw a separate flowchart to illustrate it.

4. Verify the flowchart's validity by following the path of simple test data through it.

Drawing a flowchart helps the analyst to understand the logic of a problem, since it is generally easier to see logic on paper than to visualize it mentally. Flowcharting is a way of thinking on paper. It is possible to experiment with the positions of decision and action boxes until the most logical arrangement is found. A well-drawn flowchart is a good communicating document. It makes logical interrelationships visible and shows the sequence of actions resulting from a set of conditions.

However, flowcharts have a few disadvantages. Levels of detail may easily be mixed and confused, and the flowchart itself may become long and complex. It is not always easy to trace back from a particular action to the set of conditions that caused it, and flowcharts are sometimes difficult to amend without redrawing.

6.3   Decision tables

One of the biggest blocks to effective handling of the definition, analysis, and solution of a problem is language. The procedures specialist, the systems analyst, the programmer, and the computer all use different languages. Many attempts have been made to solve this communication problem, but most solutions fail to consider that there is both a man-to-machine and a man-to-man language barrier.

Man-to-machine communication has been greatly improved by the introduction of low- and high-level programming languages. (These are considered in more detail in chapter 13.) However, the problem of man-to-man communication is immense. The most common semistandardization method is block diagramming, but even this method lacks universal standard symbols and language within the blocks.

Decision tables offer remarkable possibilities for standardized communication. A standardized decision table format with standardized decision table language provides excellent documentation for the definition, analysis, and solution of data-processing problems. Furthermore, this standardization simplifies the implementation phase of problem solution.

A decision table is one method of defining the actions to be taken when certain pre-specified conditions are fulfilled. They are often preferred for specifying actions that result from complex tests because they make possible checking that all test combinations have been considered and show alternatives side by side. They also make cause and effect relationships clear and use semistandardized language, so that unnecessary verbiage is eliminated. Typists can copy them with virtually no questions or problems, and they are easily broken down into simpler tables. They are easily understood because people are accustomed to using tables, such as train timetables, mileage tables, and so forth. Also, a program may sometimes be written directly from a decision table.

Figure 6-D shows a decision table, which is basically four quadrants divided by intersecting double lines. The table consists of the following elements:

1. *Table header,* which identifies the particular table by table number, description, and so forth;
2. *Table size C.A.R.,* which is the size of the table that is used when the table is accepted by a computer program. The code is of the form C2 A3 R5 where there are two conditions, three actions, and five rules.
3. *Rule header,* which identifies a particular rule such as rule number, type of entry code, and so forth;
4. *Rule description,* which is written vertically on the form towards the top of the table;
5. *Row header,* which identifies a particular row, usually by a row serial number of a condition or an action row;
6. *Condition stub,* which is the left part of a condition statement;
7. *Condition entry,* which is the right part of a condition statement;
8. *Action entry,* which is the right part of an action statement; and
9. *Remarks,* or any additional information.

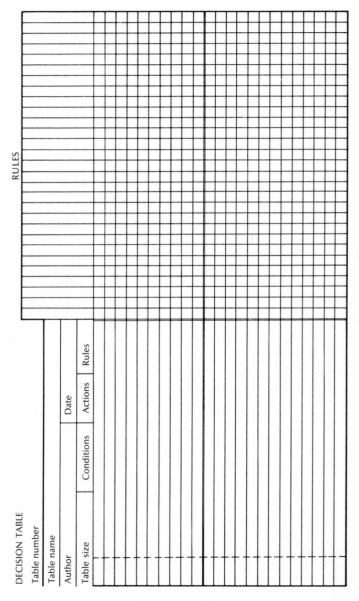

Fig. 6-D. Decision table

Tables can be completed in one of two ways: limited entry or extended entry. With a limited entry table, the statement of each condition or action is completely contained in (or limited to) the appropriate stub. The entry portions of this table indicate only if a particular rule satisfies the condition or requires the action stated. Three symbols are used in the condition entry portion of the table: "Y" if the condition is satisfied; "N" if it is not; and a dash if the condition is not pertinent to the rule. Similarly, only two symbols are allowed in the action entry portion. These are "X" if the action is to be executed, and "blank" if it is not.

In an extended entry table, the statements made in the stub portions are incomplete. Both the stub and the entry sections of any particular row in the table must be considered to decide if a condition or action is relevant to a given rule. The advantage of the extended entry method is that it saves space. An extended entry table can be converted to a limited entry form. Figure 6-E shows the two methods stating the same problem.

Limited entry tables may be checked for completeness by a simple mathematical relationship. First, count the number of blanks in each rule. Call this N, remembering that $N=0$ when there are no blanks. Then for each rule in turn, calculate the value of 2 to the power of N. Add the resulting values. This answer should equal 2 to the power of C, where C is the number of conditions given in the table. If the two values do not agree, then either some rules are missing (incomplete) or too many rules have been inserted (redundant). The table must be rechecked for accuracy. If two identical sets of conditions require different actions, the table is ambiguous. Of course, some combinations of conditions may not be realistic (as in the example, an order cannot be valued at both $0-10 and $11-100 simultaneously) and may be passed over quickly, but calculating the mathematical maximum number of rules will insure that no combinations are overlooked.

One additional feature called the *else* rule voids the above checking method. It is used when only some conditions require testing. The required conditions are written as normal rules, and a final rule having no conditions but entitled "else" is inserted. The usual actions for this rule are to go to

LIMITED ENTRY	RULES				
	1	2	3	4	5
Is order between $0–10?	Y	N	N	N	N
Is order between $11–100?	–	Y	Y	N	N
Is order over $100?	–	–	–	Y	Y
Has customer satisfactory credit limit?	–	Y	N	Y	N
Approve order	X	X		X	
Allow discount of 3%		X			
Allow discount of 5%				X	
Refer to supervisor			X		X

EXTENDED ENTRY	RULES				
	1	2	3	4	5
Order is more than	$100	$100	$10	$10	$0
Credit limit satisfactory	Y	N	Y	N	–
Approve order	X		X		X
Allow discount of	5%		3%		
Refer to supervisor		X		X	

Fig. 6-E. Limited and extended entry methods

an error routine or to exit from the table. Insertion of this special type of rule is an instruction to the programmer to perform the tests only on stated conditions. If these are unsatisfied, then the "else" action is taken, without testing every possible combination of conditions. Although this device can be useful in restricting an otherwise large table, it must be used carefully since conditions and actions are likely to be scrambled. An experienced analyst can prepare a decision table directly from the narrative without difficulty, but the following approach is suggested for the beginner:

On the narrative, underline all conditions with a solid line and all actions with a dotted line. Enter the first condition on a blank decision table outline immediately above the double line, using extended entries. Enter the first action immediately below the line, and complete the table in extended form. Identify and consolidate similar statements; check for ambiguity, redundancy, and completeness; and insert *else* rule. See if the table should be converted to limited entry by checking whether it will fit on one page, whether any existing entry will extend to more than five new lines, and whether there are a reasonable number of entries on at least two lines. Then check whether the problem would be better expressed on more than one table.

Figure 6-F shows an original narrative, its underlined version, and its decision table.

As a general rule, a decision table should be considered when the number of rules multiplied by the number of conditions is six or more. However, the tendency to draw up tables that are too large must be resisted. Another general rule is that in a full-size limited entry table, the maximum number of conditions should be four, which will generate sixteen rules. The analyst must use his common sense to detect when a table is too complex and should be split. His aim is to communicate clearly and concisely.

The analyst should also consider the programming implications of his work. The programmer codes to produce the least number of instructions. The analyst can help him design his program efficiently by stating which conditions out of those listed are expected to occur most frequently and quoting an order of importance for the remainder. Figure 6-G illustrates how this could affect processing time for a large number of transactions.

Table searches are often involved in programming decision table logic. When there are more than nine rules in the table, the binary search method described in the previous chapter may be used to advantage. For less than this number, a straightforward sequential search is usually adequate.

Another programming method applicable to decision tables is called the "rule mask" system. This method requires that the original table condition entries are set up as two binary matrices in main storage. Incoming data that has to be tested by the decision table is also converted into a binary

Original narrative

WHEN THE QUANTITY ORDERED FOR A PARTICULAR ITEM DOESN'T EXCEED THE ORDER LIMIT AND THE CREDIT APPROVAL IS OK, MOVE THE QUANTITY ORDERED AMOUNT TO THE QUANTITY SHIPPED FIELD. THEN GO TO A TABLE TO PREPARE A SHIPMENT RELEASE. OF COURSE, THERE MUST BE A SUFFICIENT QUANTITY ON HAND TO FILL THE ORDER.

WHEN THE QUANTITY ORDERED EXCEEDS THE ORDER LIMIT, GO TO A TABLE NAMED ORDER REJECT. DO THE SAME IF THE CREDIT APPROVAL IS NOT OK.

OCCASIONALLY, THE QUANTITY ORDERED DOESN'T EXCEED THE ORDER LIMIT, AND CREDIT APPROVAL IS OK, BUT THERE IS INSUFFICIENT QUANTITY ON HAND TO FILL THE ORDER. IN THIS CASE, GO TO A TABLE NAMED BACK ORDER.

UNDERLINED VERSION

WHEN THE QUANTITY ORDERED FOR A PARTICULAR ITEM DOESN'T EXCEED THE ORDER LIMIT AND THE CREDIT APPROVAL IS OK, MOVE THE QUANTITY ORDERED AMOUNT TO THE QUANTITY SHIPPED FIELD. THEN GO TO A TABLE TO PREPARE A SHIPMENT RELEASE. OF COURSE, THERE MUST BE A SUFFICIENT QUANTITY ON HAND TO FILL THE ORDER.

WHEN THE QUANTITY ORDERED EXCEEDS THE ORDER LIMIT, GO TO A TABLE NAMED ORDER REJECT. DO THE SAME IF THE CREDIT APPROVAL IS NOT OK.

OCCASIONALLY, THE QUANTITY ORDERED DOESN'T EXCEED THE ORDER LIMIT, AND CREDIT APPROVAL IS OK, BUT THERE IS INSUFFICIENT QUANTITY ON HAND TO FILL THE ORDER. IN THIS CASE, GO TO A TABLE NAMED BACK ORDER.

LIMITED ENTRY TABLE		RULE 1	RULE 2	RULE 3	RULE 4
01	Qty. ordered $\leqq$ order limit	Y	N	Y	Y
02	Credit approval = OK	Y		N	Y
03	Qty. on hand $\geqq$ qty. ordered	Y			N
04	Move qty. ordered to qty. shipped	X			
05	Go to release	X			
06	Go to order reject		X	X	
07	Go to back order				X

Fig. 6-F. Preparing a decision table from a narrative

vector. Multiplying the vector by one matrix and comparing the results column by column with the second matrix indicates whether any rule has been satisfied. The technique eliminates ambiguity in testing and allows for compactness

ORIGINAL DECISION TABLE	Rules			
	1	2	3	4
Is condition 1 satisfied?	Y	Y	Y	N
Is condition 2 satisfied?	Y	—	N	—
Is condition 3 satisfied?	N	Y	N	—
Perform action in table	A	B	C	D

Same logic as program flowchart (one possible version)

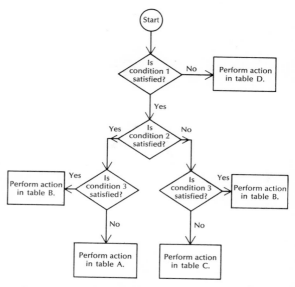

TIMING COMPARISONS DEPENDING ON TRANSACTION CONTENT

SATISFIED BY TRANSACTION	NO. OF COMPARISONS NEEDED	TIMING RESULTS			
		Per 1,000 transactions	Total comparisons	Per 1,000 transactions	Total comparisons
Condition 1	1	800	800	100	100
Condition 2	3	100	300	400	1,200
Condition 3	3	100	300	500	1,500
			1,400		2,800

Fig. 6-G. Effect of condition frequencies on timing

in object programs that result from storage of binary matrices. Its disadvantages are the need to specify every condition on incoming data to build up a vector of the required size, and the fact that common high-level languages cannot be used because they do not possess binary facilities.

A few specialized high-level languages have been devised to help programmers code decision tables directly. The two best known of these are FORTAB, which enables the decision table logic to be written so that it can be converted directly by a preprocessor program into FORTRAN, and DETAB-65, which is very similar, but translates into COBOL.

For fuller details of these programming implications, the analyst is referred to the texts mentioned in Appendix C.

6.4   The program specification

The end result of the analyst's design work is a program specification. This document provides a communications link between analyst and programmer. It must be meticulously detailed, because this communication must be perfect. The programmer depends on his specification for guidance about the innumerable questions of logic that arise as he writes his program. No loose ends can be allowed.

The programming of the system may take many man-months. By the time program testing reveals some inconsistency in the basic design, the analyst is likely to be immersed in the details of a different system. Unless his specification is clear and correctly documented, his chances of discovering the fault and suggesting a remedy will be remote. Also, his work will often take him away from the computer center into other areas of the company. If queries arise about specification while he is absent, he may not be able to resolve them by telephone. Therefore, valuable programming time may be lost while the problem awaits the analyst.

The methods of presenting the program specification may vary from one installation to another. Whatever the actual presentation method, it must be standardized. The forms presented previously are suggestions for standards. These should be supported by a narrative. The following checklist indicates details that should be included within a good program specification.

PRELIMINARY INFORMATION

1. *Title and responsibility*

    1. System name
    2. System number
    3. Date produced
    4. Produced by (name, address, telephone)
    5. Produced for (name, address, telephone)
    6. Origination of changes

2. *Authorization*

    1. Statement of acceptance
    2. Signature of person accepting
    3. Area of responsibility

3. *Contents list*

4. *Maintenance record*

    1. Change number
    2. Authority reference
    3. Date of change
    4. Section number
    5. Page number
    6. Description

5. *Time schedules*

    1. First test with live data
    2. Parallel or pilot running begins
    3. Solo running begins
    4. Other systems to be linked

6. *Definition of terms*

AIMS OF THE SYSTEM

1. *Procedures covered*

2. *Departments concerned*

3. *Adjacent procedures*

4. *Benefits*

1. Financial
2. Manpower
3. Control
4. Information availability
5. Information quality improvement

5. *Outputs*

1. Medium
2. Description
3. Uses to be made
4. Reference to specification section
5. Volume

6. *Inputs*

1. Medium
2. Description
3. Information source
4. Reference to specification section
5. Volume

7. *Files to be maintained*

1. Medium
2. Description
3. Uses to be made
4. Reference to specification section
5. Volume

DETAILED SYSTEM DESCRIPTION

1. *Clerical procedures*

1. Flowcharts
2. Narrative, including procedure descriptions (daily/weekly/monthly activities), codes, controls, distribution and handling

2. *Data preparation output distribution procedures*

   1. Flowcharts
   2. Narrative, including procedure descriptions, codes, controls, distribution and handling

3. *Computer procedures*

   1. Flowcharts
   2. Narrative, including number of runs (daily/weekly/ monthly), purpose of runs, function of programs

CHANGEOVER PROCEDURE

1. *File conversion*

   1. How data is collected
   2. Editing required
   3. Codes required
   4. Data preparation
   5. Programs required
   6. Runs required

2. *Pilot running*

   1. Data to be used
   2. Volumes to be handled
   3. Processing period
   4. Expected results

3. *Parallel running*

   1. Processing period
   2. Anticipated volumes
   3. Check points

4. *Direct changeovers*

   1. Changeover period
   2. Overtime to be worked
   3. Extra personnel to be employed
   4. Anticipated delays

EQUIPMENT USE

1. *Equipment specification*

   1. Type of computer
   2. Peripheral devices
   3. Supplementary equipment

2. *Equipment use (computer)*

   1. Number of runs
   2. Frequency of runs
   3. Setup time
   4. Estimated time per run (minimum, maximum, and average volumes)
   5. End-of-run procedure times
   6. Total computer time for daily, weekly, or monthly procedures

3. *Equipment use (supplementary equipment)*

   1. Machine type
   2. Type of activity
   3. Frequency of activity
   4. Time for each activity (minimum, maximum, and average volumes)
   5. Total equipment utilization time for daily, weekly, or monthly procedures

SOURCE DOCUMENT SPECIFICATIONS
For each document —

1. *Description*

   1. Identification (name, number, purpose)
   2. Method of origination (how and where filled in)
   3. Elements of data, including field names, description (alphanumeric/numeric), field lengths, maximum values, origin of each element
   4. Frequency

2. *Document sample*

PRINTOUT SPECIFICATIONS
For each printout —

1. *Format planning chart*

    1. Printout reference number and name
    2. System name and number
    3. Program name and number
    4. Number of print lines per sheet
    5. Maximum field sizes
    6. Horizontal spacing requirements
    7. Vertical spacing requirements

2. *Narrative*

    1. Purpose
    2. Type of form (preprinted, blank, size, quality)
    3. Special features (such as use of limited character set)
    4. Distribution and routing

3. *Samples*

FILE SPECIFICATIONS

1. *All files*

    1. Medium
    2. Name
    3. Labels
    4. Size (maximum/average)
    5. Number of record types
    6. Sequence
    7. Block size (unit of transfer)
    8. Fields per record, showing field names, field descriptions, field lengths, maximum values, permitted signs, and so forth

2. *Card files*

    1. Card types
    2. Number of card columns per field
    3. Punching code

3. *Paper tape files*

    1.   Interblock gaps
    2.   Block start and end signals
    3.   Punching code

4. *Magnetic tape files*

    1.   Tape labels
    2.   Packing density and gap size
    3.   Number of reels
    4.   Frequency of use
    5.   Retention period

5. *Direct-access files*

    1.   Storage method (random or sequential)
    2.   Storage access method (including system, address computation)
    3.   Bucket size
    4.   Number of buckets

DESCRIPTION OF SYSTEMS TEST DATA

1. *Input data*

    Listing of layouts

2. *Master files*

    Layouts

3. *Expected results*

    1.   Logic results
    2.   Arithmetic answers
    3.   Line printer results (print chart)

PROGRAM DESCRIPTIONS
For each program —

1. *Introduction*

    1.   Program name
    2.   Program number
    3.   Purpose

2. *Start procedures*

    1.   Standard
    2.   Nonstandard

3. *Main procedures*

    1.   Processing requirements
    2.   Validity checks
    3.   Control totals
    4.   Maximum, minimum sizes
    5.   Signs
    6.   Rounding requirements
    7.   Error conditions

4. *End procedures*

    1.   Output of control totals, tables, and analyses
    2.   Dumping requirements
    3.   File closing
    4.   Operator messages
    5.   Entry to other programs

5. *Dump and restart procedures*

    1.   Halt points
    2.   Dump medium
    3.   What is to be dumped

The program specification is given to the lead programmer or a senior programmer, who divides the work required to write the programs. A program control sheet, shown in Figure 6-H, is made for each program so that progress can be monitored. The control sheet names each program clearly and indicates any links with other programs within the application. It gives target dates for the program logic flowcharts, program coding, testing, and operating instructions.

When the program specification has been completed, the analyst has partially finished his task of introducing a com-

PROGRAM CONTROL SHEET

Page_____

Date_____

Program [   ]   Analyst_____   Supervisor_____

Computer system flowchart: Input/output links

Activity	Programmer assigned	Start date	Finish date	Actual days
Review specification ' Flowcharting Coding Program testing Operating instruc.				

Fig. 6-H. Program control sheet

puter system. A number of other jobs remain, however, and will be described in later chapters.

Exercises

Represent situations 1–7 by a suitable flowchart and decision table, if appropriate.

1. A document is taken from the "In" tray, read, and placed in the "Out" tray.

2. A letter is taken from the "In" tray, checked for correct date, and sent for signature.

3. A letter is picked up, read and checked for errors, and placed in the "Out" tray for collection.

4. An order is received and read. The stock number of the ordered item is checked. An inventory balance card is removed from a file, checked against the quantity ordered, and refiled.

5. An order is read; inventory data for each item is checked by reference to a stock ledger. If the item is available, the order is sent to inventory. If the item is not available, the order is stamped and placed in the pending tray until 4:00 P.M.

6. An order is taken from a file by the general manager, who signs it and sends it to the foreman. The foreman checks the raw material requirement on the order with his inventory balance card file. If he can meet all raw material requirements, he passes the order to his assistant, who files it. If there are any shortages of raw material, the foreman writes them on the order and returns it to the manager, who replaces it on file after noting the shortages on his action list.

7. The inventory control department receives a route sheet and its punched card requisitions. Inventory balance cards for the requisitions are withdrawn from a file. The

requisitions are checked against the balance cards, which are updated and refiled. Requisitions that can be filled are stamped and attached to the route sheet, which is then sent to the issue clerk. The issue clerk separates the requisitions from the route sheet, issues the parts, and stamps "cancelled" on the requisitions. He then sends the requisitions to cost accounting and places the route sheet, with the issued parts, in a tray for action by the foreman. Requisitions that cannot be filled are filed in the warehouse office. These are taken to the data-processing room once a week and processed to produce a shortage report. The requisitions are then returned to inventory control and refiled.

8. What is the difference between the two general types of flowcharts?

9. When would a decision table as a documentation technique be superior to a program flowchart?

10. What is the difference between a limited entry and an extended entry decision table?

11. Assume an order review clerk receives an order and refers to an inventory ledger card to check the balance on hand. If there is sufficient stock to fill the order, he updates the ledger with the new balance and refiles it. If the order cannot be filled, it is placed in a backorder file. Draw a flowchart to represent this procedure.

7

proving the design

Devising a system is only part of a systems analyst's job. Someone has to be persuaded to buy it and to allocate men, money, and time for installing and running it; also, its users must be able to get information from it quickly. A systems analyst has, therefore, to test and time the systems he devises.

Testing the system   7.1

The new system must be thoroughly tested to prove that it can perform its tasks. Testing is performed both by programmers and by the systems analyst.

To test a program, the analyst specifies what has to be done, and the programmer writes and tests a complete subsystem. Each program or module is tested individually and then in combination with the rest. All test data and results should be included in program documentation. When tests are complete, the programming staff should issue a set of programs that have been fully tested in main loops, error routines, and end-of-file procedures. The analyst then provides systems test data, made up to test all conditions and drawn from actual data.

The analyst performs desk checks by comparing the program flowcharts with the original specifications to satisfy himself that all conditions have been covered. He also prepares input that is known to contain errors and passes this through the set of programs. Results are checked carefully to make sure that all the errors were discovered and that relevant action was taken by the computer system or the operator.

Some common points that the analyst can test by this method in edit programs are the effects of oversize or undersize items, out-of-range items, incorrect formats, negative numbers, lack of data, invalid combinations, input of corrections, effect on daily and batch controls, batch control testing, and error bypass routines.

Input data to updating programs can be designed to examine such effects as input lacking a file record, file record lacking input, incorrect record formats, incorrect or out-of-date files, and zero and negative amounts that may cause difficulties during calculation.

The final results can be checked to test the effect of restarts because of printing or punching errors.

In systems testing, the fact that some files have not yet been constructed often causes problems. Special test files that may not contain the true record distribution have to be used. This is one of the difficulties of testing a combination of man/machine procedures that has not been fully implemented. However, the system's logic can be traced to make sure that all alternatives have been considered, by using standard documents and flowcharts.

There are two distinct types of testing: testing the system in abstract; and testing the system's operational aspects, which are the facts associated with people and machines. Figure 7-A illustrates the general steps in the introduction of a computer system and shows the two areas of testing involved. In the abstract, the test is of logic, information flow, and design. In reality, it is of the efficiency of communication and training. Abstract tests must always precede concrete tests, and they can be done in stages.

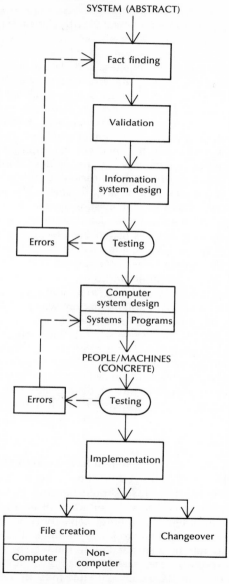

Fig. 7-A. Stages in systems testing

When the system is being designed, sub-units of program testing can be performed in parallel with clerical system testing. Eventually, testing becomes more concrete. At first form design is tested for completion and extraction. Then instructions are tested for completeness and clarity, and finally the transmission network of data and results, which involves messengers, office clerks, postal services, railways, and so forth, is tested. It is also important to test error-correction routines based on expected computer output. Response time, resubmission procedures and documents, and rejection of resubmitted data must be checked. An internal auditor's advice should be heeded, because newly-developed systems may be audited by external data-processing auditors at a later date. As a final check on accuracy, the analyst should arrange for some "live" data from the existing system to be checked in detail against manually calculated solutions, and all discrepancies should be investigated. As a result of these tests, the analyst can usually provide final proof to user management that the system does, in fact, work. Most skeptics are convinced by a computer printout that can be physically compared for accuracy against a document prepared in their own department. Then implementation, which is considered in more detail in chapter 10, can begin.

In summary, the analyst must work with programmers to make sure that their test data is effective; test the complete system; maintain documentation for management, auditors, and user departments; and be aware that program and system maintenance means further testing and test data, and that the testing process is, therefore, continuous.

7.2   Systems timing

When individual runs in a system have been timed, the analyst has to consider how they should be performed in a working day. The total load should be broken down into a monthly schedule to show the effect of peaks in processing and ascertain whether computer capacity is adequate. In a running installation, this work can be done in consultation with the operations manager. Before the regular schedule of reports or results of computer runs can be announced to individual departments, a daily schedule must be set. It should take into account such factors as time required for clerical handling and postal delays over weekends.

The operations manager must use the monthly operation schedule and the daily operation schedule to allot the jobs proposed by the systems analyst into the existing workload. Figure 7-B illustrates an operation schedule, which is basically a horizontal bar chart.

When allocating times to the jobs involved, allowance must be made for delays that may occur because of the nature of the system. This is particularly relevant when items have to be transmitted to a central preparation unit from outlying stations. Figure 7-C illustrates a typical job for preparing weekly statistical data. It demonstrates how a task that requires only one and a half days to process may require two and a half days to actually produce results.

Hardware timing   7.3

Any discussions of hardware timing must be generalized, since there are so many different types of machines. Facts about a particular machine must be provided by a timing expert at the installation, and the established timing formulas and factors should be summarized in a standards manual for everyone to use as required. The type of central processor affects both the timing and number of instructions to be executed.

In general, it should be remembered that word machines process one word at a time and are economic for large numbers (36 bits = 36,000,000,000), as their arithmetic is very fast. However, they are inefficient for small numbers. Short fields have to be unpacked and long fields may bridge word boundaries. They may require double precision arithmetic, and they use more instructions. Character machines can handle variable-length fields and work in decimal. They are economic for numbers with few digits and provide for ease of data movement, but they are slower than word machines. Byte machines provide economic storage of decimal numbers and large numbers of special characters. They can handle variable-length fields and work in decimal or binary. They can use registers and can also work from storage to storage. The analyst should also remember that the larger the number of

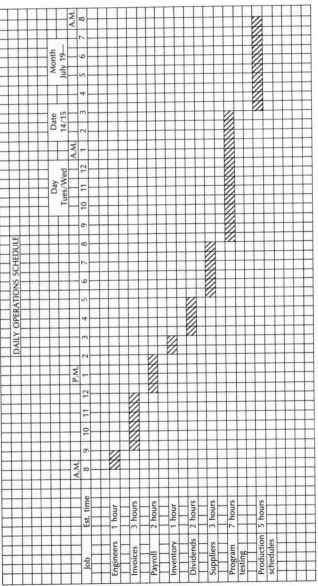

Fig. 7-B. Daily operations schedule

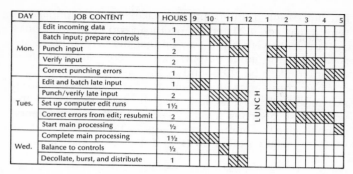

DAY	JOB CONTENT	HOURS	9	10	11	12	1	2	3	4	5
Mon.	Edit incoming data	1	▨								
	Batch input; prepare controls	1		▨							
	Punch input	2			▨	▨					
	Verify input	2					▨	▨			
	Correct punching errors	1									▨
Tues.	Edit and batch late input	1	▨								
	Punch/verify late input	2		▨	▨						
	Set up computer edit runs	1½				LUNCH		▨			
	Correct errors from edit; resubmit	2							▨	▨	
	Start main processing	½									▨
Wed.	Complete main processing	1½	▨	▨							
	Balance to controls	½			▨						
	Decollate, burst, and distribute	1				▨					

Fig. 7-C. Timing example for results production

available instruction capabilities, the fewer instructions required to solve a problem.

In order to time the internal processing of a computer run, it is necessary to know the average instruction time for an average commercial problem and the number of instructions to be executed. Only the main loop in the program need be considered; an adjustment can be made for exceptions afterwards if desired. Process time equals number of instructions in main loop times average instruction time in microseconds.

In timing input and output units, it is necessary to be familiar with the operating system, which may be able to read in or write out information at the same time as processing. The types of operating systems available include the following features:

1. No overlap;
2. Simultaneous input and output, not overlapped with processing;
3. Buffered units that permit some overlap and only have to wait for data transfer from a fast buffer;
4. Use of several buffers in main storage so that complete overlap is possible.

However, the job can never be run faster than the slowest unit. Other units will have to be delayed temporarily at intervals to await completion of operations involving this unit.

The computer may also have an interrupt system so that processing can continue regardless of the input and output, but processing of interrupts also takes time. In simple systems it may add a 20 percent overhead and in complex ones a 50 percent overhead. It may be that the central processor is fast and the input/output units are comparatively slow, causing runs to be I/0 bound in any case.

When considering the input of data to the system from magnetic media, factors other than the working speed of the units must be considered. The manufacturer's reference manuals give this information. Figure 7-D summarizes formulas for calculating times on tape, disk, drums, and cards.

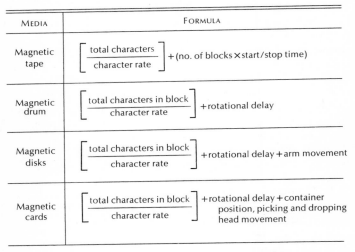

Fig. 7-D. Generalized timing formulas for magnetic devices

Peripheral devices may be connected to channels and their characteristics may affect timing. If many tapes are on one high-speed (selector) channel, only one tape can be used at one time. Several card readers, punches, or printers can be used at once on a special slow speed (multiplexor) channel. Systems using high-speed channels may be limited by numbers of channels; two devices cannot read to or write from one channel at one time.

Peripheral devices can do some actions independently, such as arm movement, tape backspace, and rewind. Only data movement ties up the channel. Channels are, in fact, small computers in their own right with their own rather limited instruction sets.

It is possible to have several input/output areas for one device so that record after record can be read as a buffer against uneven processing time. Normally, two areas per file are reserved. The device can carry out a sequence of instructions without any interruption of main processing. Data is placed in core storage or taken from core storage in competition with the main processing program, which also requires the storage mechanisms. This is done by "cycle stealing," in which processing of the main program stops for one core storage cycle and then continues. If this method is used, the width of the data path and the core cycle time must be known. The required timing formula is then

$$\text{Time lost to main program} = \frac{\text{Total number of bits} \times \text{cycle time}}{\text{Width of data path}}$$

To calculate the overall run time, it is necessary to time each component. Those that have the longest times determine the run time. Obviously, the best way to calculate run time is to process a measured batch of typical test data through the system, time it precisely, then extrapolate.

In some cases, references to a file may be unevenly distributed. It is necessary to know the percentage and frequency of file activity. Alternative systems approaches should be considered to avoid passing a tape file unnecessarily often, or having very large disk files online continuously when only a small fraction of records is normally required.

Reference to a disk file may be by:

1. Direct addressing—imposing limitations on the use of reference numbers;
2. Indirect addressing by randomizing—using a formula to convert actual references to addresses; or
3. Using an index for each cylinder.

Using an index involves additional disk reading time whenever reference is made to a different cylinder. The time to read the first record on a different cylinder is given by the following formula:

Time = Arm Movement + Rotational Delay + Reading Index
              + Rotational Delay + Reading Record

Reference to a record on the same cylinder does not usually require reading the index into main storage from the cylinder. It is usually still available in main storage.

When programs are stored on disk by the operating system, it may be possible to call in special subroutines from a library on disk to handle unusual conditions. These subroutines should be planned to fit into a common transient area or areas, and to allow execution of one subroutine to begin while others are being read in. Use of subroutines should be the exception rather than the rule, so that run time is not significantly increased.

Many of the topics discussed in this section are relevant to program planning, which is normally the responsibility of a senior programmer rather than of the analyst. This brief outline has been given so that the analyst can discuss modifications to improve the system's timing performance.

Exercises

1. In designing systems test data, what types of conditions does a systems analyst consider?

2. Discuss this statement: "The best way to test a system is to run a batch of live data through it."

3. What advantage is there in having the systems analyst, rather than the programmer, prepare program test data?

4. Why is it important to have a well-defined schedule for individual runs in a system?

5. What is the most efficient way to estimate the time required for a given program to process a given amount of data?

communications

During all stages of his work, the systems analyst must communicate his ideas and designs to others both formally and informally. Progress reports and completed systems specifications are examples of formal methods. Informal communication can be achieved by sharing charts with other analysts or having off-the-record discussions with operational staff, for example.

The analyst must be able to communicate well if he is to convince others that his ideas are practical. He must be able to express himself clearly both orally and in writing. He may have to explain his system to audiences of widely differing backgrounds, ranging from senior management to shop floor workers. He must also be able to prepare concise reports and any other necessary documentary support. The following sections outline the main items the analyst needs to understand to communicate effectively.

8.1   Communicating the system

The introduction of any new system will have far-reaching effects if it integrates work flows. It will affect the future work pattern in user departments, change their functions, and often alter the responsibilities of their staff. Some existing departments may disappear, and new ones may be created. In such an atmosphere of change, the system will not work unless both its details and the method of implementation are properly communicated throughout the organization.

Much of this task can be handled through an implementation committee whose members are drawn from the departments concerned. The committee can centralize communication between departments carrying out the changeover, user departments that are directly affected, and senior and middle management who are indirectly affected.

The total communication task will incorporate education, training, reporting, and control. The introduction of a new system will interrupt daily tasks and lead to insecurity. Even persons not directly affected may be concerned about future extensions of computer systems. Employees should be given information about the new system and how it affects them, so that they can understand the system in the context of the organization. This can be done, for example, by printing articles in staff magazines, organizing visits to the computer center, holding discussion meetings where questions are answered by the implementation team, and keeping union representatives well informed.

Training consists of teaching people new skills, or showing them how to do a job in a new way. Training facilities must be available, and newly acquired skills must be used as soon as possible. A comprehensive training program requires handbooks and manuals, lectures and demonstrations, examples of new documentation, and visual aids.

However, education and training are not one-way channels. For example, difficulties in teaching may reveal defects in form design. There must be feedback as the system adapts itself. Feedback is best summarized in formal reports. Discussions should be documented whenever possible and if a discussion involves more than one department, the analyst should remember that several reports may be required.

Reporting and control work together, but establishing effective control cycles may be difficult. If the implementation committee meets at irregular intervals, predetermined control cycles may not coincide with committee meetings. Under these circumstances, the committee will not be able to review progress and exercise effective control. The full committee must then delegate its authority to one member whose task is to make sure that the required control function is executed at the proper time. The obvious candidate for the task is the systems analyst. For this reason, he will often find himself in the position of secretary of the implementation committee. As such, he will be the focal point in the communications network stemming from his system. His aptitude for the job is obviously a vital factor in the overall success of the implementation.

Audio-visual methods of presentation   8.2

The systems analyst presenting his new procedures to management or the operational staff must express himself well. A clear and concise exposition of the new system will help overcome natural resistance to change and enable new procedures to be implemented with a minimum of disorganization, by creating fuller understanding.

Perhaps the most common way in which the analyst will communicate his ideas is by a formal lecture, usually to a few of the staff. The most important aspect of this task is adequate preparation. Preparation is easier if a general framework is followed. A suggested framework is:

1. State the proposal;
2. Concede the objections;
3. Support the proposal with one main argument;
4. Give evidence to support the argument;
5. Restate the proposal; and
6. Ask for questions.

A lecture is often more interesting and explains points more clearly if visual aids are used. Various methods are available to the analyst. Films are usually very good as an introduction to EDP, but they are expensive to make and are usually best for general topics rather than specific problems.

Using an overhead projector may be useful, since it is easy to produce reusable transparencies, and diagrams can be built by laying one transparency over another. With a chalkboard, diagrams can be built and altered quickly. But once erased, information on a board cannot be referenced without redrawing. Flannel boards and magnetic boards both allow for diagrams to be built, and they are more portable and less messy than chalkboards. However, the representation of the information on them must be constructed ahead of time. Flip-charts are easily transportable and usually cheaper to produce than slides, but cannot be reused if drawn on.

8.3   Report writing

Reports are formal communications of the reasons for, the nature of, the results of, and the conclusions from a particular action or investigation. The purpose of a report and how it will be used should be made clear before it is written, so that the person who writes it can supply all the relevant information.

Reports that are intended to persuade people to act should be oriented towards the recipient. This is not to say that facts need not be true or that opinions need not be honest, but that the report and language should be structured to interest the reader. It is not justifiable to state partial truths or to be vague about things that can be precisely stated, such as saying "a high percentage" instead of 51 percent, or "a survey was made" when the writer talked to 3 of 10 workers.

The introduction of a report should explain why the report is being written and define its limits. Any definitions or qualifying facts necessary for understanding should be noted.

The main body of the report should be divided into sections and presented in logical sequence. In order to maintain interest, detailed procedures, specifications, charts, and so forth, should not interrupt the main trend of the report. They should be placed in appendixes and briefly referenced in the main body.

There are three main types of information used in reports. Descriptive or factual information discusses facts and may also describe inferences that can be drawn from them. (However, if the validity of the inferences is discussed, the information is no longer merely descriptive.) Instructional information shows how to do something or what to do, and can range from arbitrarily imposed edicts to suggested advice that is supported by convincing arguments. Dialectical information presents opinions and ideas based on logical inferences from a series of definitions or observations and states reasons for these opinions.

Semantics—that is, word meanings—should also be considered. Words may have "private" meanings, new words may not be universally understood, and the meanings of words may change over time. If in doubt, a recent dictionary should be consulted. Words that refer to specific facts and ideas rather than to abstract generalizations are usually preferable for report writing.

Following is a brief outline of items to consider when writing a report.

Defining task—terms of reference

1. What is being written about?
2. Why is the report needed?
3. What effect may the report have?
4. Who are the readers and what do they know?
5. What is wanted—guidance or firm proposals?
6. What work has been done previously?
7. What other reports have been made and what were their conclusions?

Information contained in the report

1. What time period does the report cover?
2. How old is the report's information?
3. What is the source of each quoted item of information?
4. What procedure will be followed to check the quoted information?

Method of preparation

1. Which person or group is responsible for preparing and writing the report?
2. How long will it take to prepare?
3. How is the data in the report compiled?
4. How many copies are to be prepared and to whom are they to be sent?

Effectiveness of report

1. What is the report designed to accomplish?
2. What use will each recipient make of the report?
3. Does the report meet all its requirements?

8.4 Data-processing department reports

The previous section considered basic principles that apply to the presentation of any written report. Some special-purpose reports are generated by the data-processing department to summarize and convey the results of the department's work. They are prepared for three levels:

1. For senior management, to give advice on overall implications and advantages of new computer applications, and on the general running and efficiency of the department;
2. For middle and lower management, to indicate the effect of new and revised procedures on their departments; and
3. For use within the data-processing department—that is, systems specifications.

Reports to senior and lower management about the adoption and implementation of new systems are the main concern of this section. Descriptions of systems for detailed implementation, reports on departmental efficiency, and detailed systems specifications to programmers are considered elsewhere.

The systems analyst's involvement in preparing these reports depends on his seniority. As a junior, he will do much of the preliminary investigation and details of design, but usually he will not prepare the reports. As a senior, he may write the major parts of reports to middle and lower management. The reports to senior management may be largely the work and responsibility of the data-processing manager, chief systems analyst, or project leader.

When recommending a system, the analyst should consider it from the viewpoint of the management system, the computer system, the human aspects, the operational system, implementation, costs, and the design process. He should give a non-computer exposition of the system, relating it to the management environment in which it is to operate as well as a computer-oriented discussion of the way in which the computer is to be used, with particular attention to the file structure of the system. He should describe the human functions involved at all levels of participation in the system, with particular attention to the detailed design and use of input and output forms and devices. The operational aspects, including hardware, with particular attention to timing and efficiency, should also be described. Further, an assessment of the quantity and quality of resources (machines and people) necessary to meet the project schedule, and a presentation of the estimated costings for all phases of the project should be made. He should give an account of the work of problem identification, problem solution, and system design. Whenever possible, the presentation should identify points at which the proposed design is sensitive, or at which special measures have been taken.

Senior management will receive what is called a feasibility report. This is a broad study of an application area with particular reference to the suitability of data-processing techniques. It also considers the cost of instituting data-processing procedures, with regard to the existing workload (system, programming, and machine) of the data-processing department, and the cost and impact of the acquisition of any additional required computing resources. These reports usually originate from manufacturers (generally as "proposals" for equipment), consultants (particularly if the company has no DP department or wants an independent evaluation of manufacturers' proposals), DP managers (particularly in smaller installations), or project leaders. If acceptable to higher management, the report may initiate the detailed systems study, which generates the systems proposal.

The systems proposal is the most important document the systems analyst produces in his function as "salesman" for data processing in his organization. Its purpose is to obtain management's approval by convincing them that the proposed system will meet all the major requirements economically. Once approved at this level, the detailed specification to programmers may be completed. His report should assure management that the system will give them what they want and show how they will benefit from its implementation.

The accuracy of all the figures, particularly file sizes and machine loadings, must be validated. The report should also show that the system is stable and will require as few changes as possible in staff. It should persuade management that their cooperation is important and emphasize the need to minimize changes to the system, particularly to outputs, during implementation.

Exercises

1. Prepare a five-minute lecture on any DP subject, using the framework presented in this chapter.

2. What types of visual aids would you use, and why, if you were presenting the following talks to the audiences shown?

Title of talk	Audience
"The new order processing methods"	Clerical staff in affected department
"Computers in management"	Board of directors
"Sales forecasting techniques"	Middle sales management
"Error detection with check digits"	Ledger clerks
"Future pay notification system"	Assembly shop workers
"Simulation using computers"	Senior production management

3. Write a feasibility study to justify the installation of a computer configuration to handle the following tasks: payroll, sales accounting, invoicing, and marketing statistics.

4. List the contents, with a brief indication of the subject matter in each section, of a systems specification detailing an inventory control system.

5. Draw up a timetable for a two-day seminar (non-residential) designed to introduce senior management to the future computer plans for their organization.

6. Why is it important for those people not directly connected with the computer installation to understand the
   system?

7. Discuss briefly the advantages of various audio-visual
   aids.

8. What points should a systems analyst consider when
   writing a report on his investigations and conclusions?

9. What are the three levels of people within a company
   to whom data-processing department reports are usually
   directed?

justifying the system

Making sure that the computer system justifies itself by achieving desired results is often forgotten. In many cases, if the job runs correctly on the computer, no one notices that large sums of money are spent without adequate return, inefficient personnel remain inefficient, and few standards of procedure or performance exist.

Obviously, procedures and time schedules must be planned. Employees must understand standard procedures and the level of performance expected of them, but this requires much more understanding of basic problems than is usually believed. Many people are involved on the fringe of computer applications, but few are released from normal duties for the necessary training.

An implementation plan must have objectives. If they are not met, adequate feedback must be provided so that they can be revised if necessary. For example, if the systems analysis for one application is behind schedule, a thorough check must be made to see whether the cause of the delay will affect other systems studies.

Often using a computer is justified by the fact that the organization's systems objectives are met. People often learn to live with, and accept, an inaccurate system simply because it is not their problem or they do not know how else jobs could be done. The speed, accuracy, and checking techniques of the computer system should lead toward objectives being met more accurately.

Defining computer benefits   9.1

Some people in any organization will always say that their system was much better before "something" happened. When the "something" is the introduction of a computer, such people will revel in pointing out errors that have occurred as a direct result of computer processing. The analyst must be able to counter this skepticism with sound arguments about the benefits of the intelligent application of computers within that system.

However, the computer salesman's hint of saving personnel salaries is rarely justification for installing a computer system. In a few cases, substantial manpower savings are made; in others, such savings are offset by the need to recruit more expensive specialists. However, once the computer system has the resources to function successfully, it is likely to be capable of absorbing more work as the business grows, without increases in personnel.

A more significant benefit of computer systems is that they enable the organization's financial resources to be used more effectively. For example, an efficient inventory control system will require less capital and yet give more satisfactory results. Similarly, investment in a project may be more carefully controlled, since resources can be purchased when they are required. A business may be unwittingly carrying unprofitable lines merely because insufficient accurate data is available. This data may be produced as a by-product of computer systems.

Growing organizations that want to continue to grow often use new equipment and techniques to obtain an advantage over competitors. For example, many businesses bid for work. In the past, a good estimating staff knew when they obtained a very profitable contract. However, when the profit margin was relatively small, they could not estimate accurately and would in fact take some contracts at a profit and others at a loss. They many variables and their fluctuations led to inaccuracy, since a manual system could cover only the most frequent or most important variables. This is an area where the computer can provide valuable aid. With its help, any number of variables and complex estimating procedures can be calculated quickly and accurately. Changes calling for recalculation of estimates can be serviced quickly.

The use of computers will eliminate the need for "hunch management." It has been said that an acceptable business manager is one who makes more correct decisions than wrong decisions. If a computer can help to eliminate wrong decisions, business effectiveness must increase considerably. The speed of information provided by computer is usually the greatest single factor in improving effectiveness through the management function.

A business reacts slowly to external changes. It is as if the business were determined to follow its original plans and resist all internal changes until the original plans become impossible to follow. The introduction of a computer system can sometimes bring about a remarkable psychological change through such reasoning as "the computer says this is now impossible—and we must act accordingly." Substitute the name of a person for 'the computer' and the same information will probably seem unimportant. A fact given by a person appears to be an opinion to others. A computer that highlights important exceptions and deviations quickly will markedly increase a business' effectiveness.

Few systems analysts have failed to notice that substantial benefits to the company arise from the computer investigation itself, even if the proposal is not implemented. Most businesses grow like Topsy. Inefficiencies are built in and treated as part of the essential fabric of the system. For instance, records are duplicated, reports that ceased to have value years ago are produced, and information is sent to managers much later than necessary because of some impediment long since removed. Few businesses spare the time and money for a systems analysis to detect these inefficiencies unless a computer is contemplated.

At some stage, an estimate must be made of the overall cost of the new system. Unless this is done, management has no firm basis for comparing the proposed system to existing methods. Intangible costs associated with "better management information," "faster customer service," and so forth, are difficult to quantify in cash terms. However, a surprising number of positive expenses are associated with any computer installation. The analyst must know the extent of these if he is to provide information as a basis for calculating personnel cost, overhead, and operating costs.

No generalizations can be made about personnel cost, except to say that trained specialists are expensive. Figure 9-A shows personnel costs of a well-established installation renting a large computer. The total salary figure should be increased by a percentage to allow for such items as taxes and pension contributions. Personnel cost is very significant in any installation and can often exceed the annual rental for the computer.

After a suitable cost per hour for the systems and programming activities has been established, the staff time required for fact-finding, analysis, design and planning, program coding, testing, file conversion, and implementation duties must be estimated.

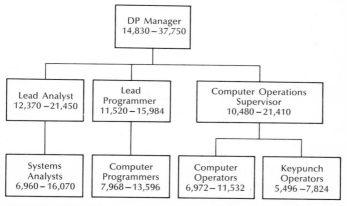

Computer World National Survey by Source EDP, 1970 edition, and *Computer Salary Survey*

Fig. 9-A. Annual personnel salaries

Fact-finding and analysis costs depend on the depth of the study, which can be anticipated only on the basis of enlightened guesswork. In this difficult area, the practices of the better consultants, whose systems are usually costed accurately, are worth noting. Sometimes their costing is accurate because of their extensive and varied experience, but another important factor is that they define their project parameters and terms of reference closely and that they do not permit changes without written notice. Often in fact-finding, the analyst is diverted to other areas, given wider terms of reference, or delayed while awaiting management backing. In such cases, original cost estimates are useless.

Costs of programming vary enormously, and performance standards are only just now emerging. A normal programmer workload produces eight to ten coded instructions per day. Many computer managers agree with this figure, but feel that it can be improved.

The cost of converting existing files to the chosen medium is a one-time expense that often occurs during the immediate pre- or post-installation period. The manufacturer can often estimate the time involved, which can be, in turn, converted to cost. If the file is set up by an external agency, an estimate is usually available. Since the work can be performed by experts, it may be cheaper. However, many installations use slack time available after they have acquired their computer for this task.

Overhead includes an amount to recover the initial costs of the site and covers such items as accommodation for the computer, engineers, and operators, and analysis, programming, and control staff. The computer room usually has the highest initial cost because it needs air conditioning, false floors, fire detection devices, soundproofing, etc. Often, accurate costing is difficult because the computer occupies one part of an administrative area and specific figures are unobtainable.

Data-preparation and related office equipment are other items of overhead whose cost must be recovered. Their cost varies from installation to installation. The average capital cost of punched-card preparation rooms is higher than the cost of paper tape installations, but this is often because some old equipment is left from the previous punched-card machine room. The cost is also affected by the size of the installation, and the applications involved. The analyst should take a serious interest in what equipment is specified.

Running costs, other than for personnel, include all the supplementary items necessary for the installation to become a working unit, such as the computer consumables, including printer forms, punched cards, and magnetic tapes. Control on these is often inadequate, and waste often occurs. For example, multi-part plain paper is often used for single-sheet listing jobs, or files are written on 2400-foot magnetic tapes when 600-foot tape would be sufficient. Also, special pre-printed forms are often prepared for internal reports that could be produced on plain paper with computer-printed headings. Moreover, these costs are frequently ignored in costing exercises, although they can involve expenditures of several thousand dollars annually.

Maintenance charges are another recurring expense. They are usually the subject of a contract with the manufacturer, usually on a one- to five-year basis. After that, maintenance must be renegotiated. There is a wide range of suppliers and equipment for data-preparation work; hence, generalizations about cost are impossible.

Costs of maintenance of the room itself and air conditioning must also be calculated. Power charges can also be considerable, depending on usage.

The above cost areas are not intended to be exhaustive, but are suggested as guidelines. Some miscellaneous charges for training courses, manuals, publications, and so forth may be added. In any particular installation, the analyst should be able to draw a comprehensive list to help in costing the system. How far he will be involved in this area of the work is uncertain; the task is sometimes undertaken by the data-processing manager. However, even if he is not concerned with the overall costs, the analyst must be prepared to obtain costs for and justify those parts of the system that incur additional expense, such as special forms and extra data-preparation equipment. In presenting this information, it is useful to understand the framework into which these costs must be inserted.

Exercises

1. Why is it important to establish performance standards and to measure a system's efficiency in terms of costs for benefits received?

2. In general, what types of benefits accrue from the systems study and installation of a computer system? Is there any way to assign a dollar value to these?

3. In determining the total cost of operating a data-processing installation, what cost factors should be considered?

4. In what ways could you measure a programmer's production, or output?

5. How would you measure a systems analyst's production, or output? Is it feasible to compare one systems analyst's production with another's?

implementation

An important part of the systems analyst's job is to make sure that his new methods are implemented successfully. This task requires the ability to coordinate and organize. The analyst must insure that the many non-computer parts of his system work effectively and that valuable machine time is not lost through manual errors caused by poor implementation.

Implementing a system requires a thorough knowledge of the design that only the analyst responsible for creating the system can have. But when a complex system requires implementation, the analyst must delegate some of the detailed work to others. To do this and yet retain firm control over the progress of the implementation requires planning.

Since employees outside the computer area will be involved in the implementation of the system, the analyst will need to establish a training program. To reinforce new skills, employees must be provided with handbooks and job aids—another facet of the overall job of implementing the system.

Almost certainly, with any new commercial computer system, files will need to be created, and this task will need careful supervision. Finally, the analyst must consider the problems that arise when the changeover is effected and be prepared to act quickly if something goes wrong.

All these varied aspects can be thought of as part of the implementation process, and will be discussed in this chapter.

10.1   The management of implementation

Implementation is the practical job of putting a theoretical design into practice. It may involve a complete computer system or the introduction of one small subsystem. A number of people, mainly within the computer department, will have been concerned with the system during its design stage. When implementation time arrives, employees in the user area will become involved. The usual approach is to appoint a coordination committee to make sure that implementation is carried out smoothly. The systems analyst is usually responsible for this committee and often acts as its secretary.

In establishing the committee, the analyst must consider the size of the job, its complexity, influences within the organization, costs and savings involved, and knowledge of both the old and the new systems. If the committee is to function well, all members must work as a team. Responsibilities must be delegated to achieve desired results. The analyst must use tact while making sure that the work is divided in such a way that the job will get done. A schedule is vital, and regular progress meetings should make sure that it is being followed. It is disastrous to have non-computer personnel provide information by a certain deadline, only to find that the required programs are not working at that time.

Figure 10-A shows the main organization features involved in implementing a system.

10.2   Scheduling the tasks

Planning is deciding what is to be achieved and how it will be done. It requires logical thinking and precision in defining actions and results. Two common aids to planning are charts and networks. The most common type of chart is the hori-

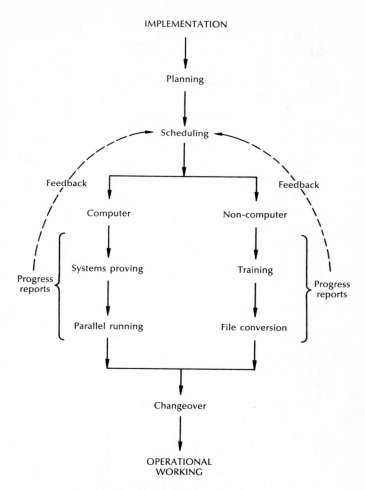

Fig. 10-A. Implementation structure

zontal bar chart, which is sometimes called a Gantt Chart, as in Figure 10-B. Activities are listed in the left column, and a time scale is drawn across the top. An open bar indicates the estimated length of time required for each activity. The bar is positioned below the weeks on the time scale that correspond to the weeks for the activity. As activities are

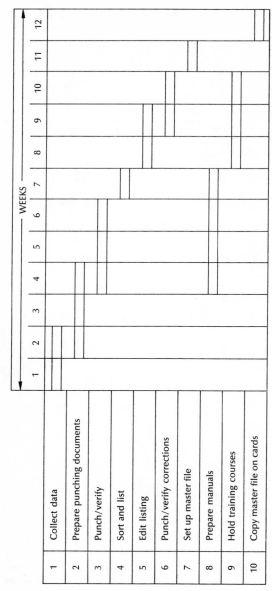

Fig. 10-B. Horizontal bar chart

completed, the appropriate bars are shaded. A transparent ruler or a piece of cord is placed in the current data position. Activities that are behind schedule can be easily seen from the unshaded portions of the bars. Squared paper simplifies construction of such a chart and makes reference to the individual entries easier.

The basic disadvantage of charts is that they cannot show interrelationships and interdependence of tasks. Networks overcome this disadvantage since their function *is* to show interrelationships, although the time scale is abandoned. A mixed system may be used with a network for planning, scheduling, and overall control, and charts and graphs for control of small sections of the task.

Figure 10-C illustrates a typical network diagram. The following definitions are used in discussing network diagrams.

1. Activity—the application of time and resources that are needed to progress from one event to the next;
2. Event—a specified objective in the overall plan to be achieved at a particular instant in time;
3. Critical path—the path traced through those activities in the network which together constitute the longest overall time;
4. Float—the excess time that can be added to any activities not on the critical path, without altering the overall completion date of the project.

The first step in drawing a network diagram is to list all events in the plan and number them sequentially for reference. The numbers of events immediately preceding and following each event and a realistic estimate of the time required for the particular activity are needed. Then events are plotted on the diagram by numbered circles and joined by arrows. The length of the arrows has no significance, but an activity time is noted on each arrow.

By tracing every possible path through the diagram from start to finish, a total time can be estimated for each one. The path with the longest time is the critical path; it has no float. All other paths contain a float, which is equal to the difference between their overall time estimate and that for the critical path. A start and finish date can be established for the overall project and its components. If events on the critical path are completed on schedule and if delays on non-

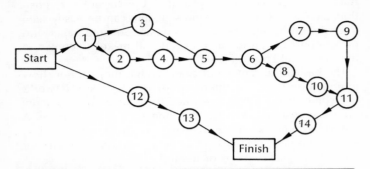

| Event No. | Description | Event No. | | Weeks to Complete |
		After	Prior	
1	Collect 50% of data	S	3	1
2	Collect balance of data	1	4	1
3	Prepare punching documents for first 50%	1	5	1½
4	Prepare punching documents for balance	2	5	1½
5	Punch and verify	3/4	6	3
6	Sort and list	5	7	1
7	Edit listing—first part	6	9	1
8	Edit listing—balance to first part	6	10	1
9	Punch and verify corrections	7	11	1
10	Punch and verify corrections to balance	8	11	1
11	Set up master file on magnetic tape	9/10	14	1
12	Prepare clerical operating manuals	S	13	4
13	Hold training course	12	F	3
14	Copy master file on to individual record cards	11	F	1

S = start   F = finish

Fig. 10-C. Network diagram

critical paths do not exceed their available float, the project will be completed on time. A tight monitoring system can be established over the progress of the whole plan; this is one of the benefits of network planning.

When the number of events and activities in a particular plan is small, manual analysis of the network is feasible. However, when a network is large, evaluation of alternatives is tedious and prone to error. It is advisable to use a computer to perform the necessary calculations. Many standard applications packages—such as PERT, meaning Program Evaluation and Review Technique—exist for this purpose.

Since its introduction in 1957, network planning as a general management technique for project control has reached an advanced state. Many standard texts on the subject provide a fuller appraisal of its possibilities.

Networks offer many advantages. Among them are that they show relationships clearly and establish an overall time limit for the task. Each job must be examined and defined in great detail to create discrete tasks so that resources can be distributed. These can be allocated to meet a projected completion date, or the completion date can be reprojected, based on current availability of resources. However, it should be remembered that networks are not drawn to scale, that the chart *itself* cannot record progress, and that visually the current position of each activity is not obvious.

Once the network is constructed and the critical path has been established, it is a simple task to convert it into a bar chart for progress control. All activities are listed and shown as individual bars on the chart, and activity times determine bar lengths. Event numbers appear on consecutive rows. If available resources are converted to a similar time scale, they can be superimposed on the basic chart. Any float time can be used to adjust activities to resources. The completed chart, combining activities and resources, can then be used to plan recruitment and training.

The following points should be remembered during planning and implementation.

1. Think precisely and reduce implementation to a series of small manageable tasks for which you can "guesstimate" a time target.

2. Check your intermediate target dates — are you in front or behind? — will this affect the final target date?
3. Avoid continually revising your dates, or people will think you can't plan at all!
4. Make sure everyone is involved in planning your targets and in working to meet them.

10.3   Training, handbooks, and job aids

Resource allocation, as shown on the networks and charts, will determine the personnel requirements at various times in the implementation cycle. Realistic personnel estimates should be made and recruitment and training must be considered. Staff requirements will vary between the creation and changeover periods when the system is being prepared for operation, and will vary again at the operation stage. In the final phase, some employees may be transferred from departments whose functions have been superseded by the computer system. For example, former clerks may become control clerks at the computer center when an invoicing application is introduced and the sales ledger is maintained on magnetic tape. The personnel department is usually responsible for recruitment. The systems analyst must be prepared to assist it with advice about the specific skills required for new or changed tasks. With his detailed knowledge of the system functions, he can often identify suitable job candidates.

If the attempt to match available employees with new jobs fails to complete the recruitment process and some jobs remain vacant, decisions must be made. Can current employees be retrained? Should trained personnel be recruited? Should untrained personnel be recruited? To answer such questions, the analyst/personnel team must consider the probable effects of delay caused by training and assess training effectiveness. If outside recruitment is to be considered, the effects of delays caused by advertising, interviews, and selection must be estimated. The advice and experience of the personnel department will be valuable in the latter situation.

If training is necessary, proper schedules must be set up. The time required to prepare courses must be taken into account. Once a course has been planned, arrangements are necessary to release personnel from their current jobs to attend it, and to continue training under practical conditions after it is completed. Obviously, trained personnel will be needed as soon as implementation begins, so training must begin much earlier. It is often advisable to appoint a member of the personnel department to the implementation committee to make sure these factors are given proper consideration.

Standards must be set to measure the effectiveness of the training program and to prepare for the possibility of failures. Reports from independent assessors will help. Monitoring on-the-job training is difficult, because reports are subjective. Appraisal of post-training reports should be approached from the point of view that if a failure has occurred, the following questions are relevant—

1. Is it the system's fault; has operation revealed a defect or weakness?
2. Is the job right; should it be altered?
3. Is the trainee wrong; if so, is it a defect in training, or a defect in the employee selection process?

With this approach, one can insure that if an employee is dropped from the program, he can be correctly replaced.

During creation and changeover, the new methods that are introduced must be defined clearly, so that men and machines can be told precisely what to do. Programming tells the machine; the system must tell the men. Whereas computer instructions have to be re-presented each time, because the computer has no retentive memory, humans learn and remember. But human memory is fallible, and it is sometimes necessary to present memory aids such as a handbook that describes in detail how the job is to be done, or a job aid, which assists a person in carrying out instructions while *performing* the job.

Within a data-processing system, the following types of handbook are used. The parentheses indicate the persons usually responsible for their creation.

1. Computer operating manuals (programmers)
2. Software manuals (manufacturers)

    3. Systems specifications (systems analysts)
    4. Program specifications (chief/senior programmer)
    5. Punching instructions (operating manager/systems analyst)
    6. Clerical procedure manuals (systems analyst)

These handbooks are aimed at different audiences, and the level of presentation within each will vary accordingly. Some should be arranged so that specific sections can be abstracted for issue to affected personnel. This type of arrangement requires that sections be self-contained and not include extensive cross references.

Standardization of manuals is impossible because of the disparity of their contents, but it is necessary within related areas such as in software manuals and program specifications. Certain standards can also be maintained, such as using consistent paper sizes, binder colors, and separator sheets.

A job aid can help in implementation of the new system. It can take innumerable forms, some of which are listed below.

    1. Use of headings and boxes in forms to emphasize points; color printing.
    2. Wall charts showing processing run schematics or large diagrams of forms.
    3. Notices placed in a position where they can be related to a certain task, such as switch positions.
    4. Use of color to identify pieces of equipment, pipes, and cables; colored warning lights and status lights.
    5. Use of various paper tape colors to distinguish between verified data, raw data, control messages, programs, and computer output; cards edged with colors for identifying types.
    6. Use of colored forms with matching colored pockets in sorter.
    7. Use of flowcharts and decision tables for error handling; operating instructions for programs with specific actions required at each stage.

Both handbooks and job aids are aids to communication. One must consider *with whom, why,* and *when* the user will be communicating before they are designed. These three points will prevent the following fictitious example from becoming reality.

A bell rings in a computer room. The operator looks up "Bell" in the operating instructions handbook and reads:

> If ringing emanates from a bell painted red, mounted on the wall to the rear of the operator when facing the console, this indicates that the temperature and/or smoke alarms have been activated. These alarms will only be activated when the temperature is 90 percent and/or smoke density is 1-2 percent of normal. In such event refer to Emergency Manual EM 693/5, latest edition. Sect. 7, para. 5 (b).

Finding the manual and the section, the operator then reads:

> The activation of the alarm system as described in EM 693/5 Sect. 3—BELL indicates the presence of a fire. In order to extinguish this and to prevent the spread of the fire, one minute after the alarms are activated and the bell begins to ring, the automatic carbon dioxide system will come into operation and after a further two minutes (during which time the bell will keep ringing), the doors to the computer room, maintenance area, and supply room, all within the area protected by the carbon dioxide system, will be automatically locked. Anyone who has to remain in the area will have to wear breathing apparatus (for location of such apparatus and for method of use see EM Sect. 5, paras. 3 and 9), but it is recommended that no personnel remain in the protected areas. Personnel should be evacuated as soon as possible after the bell has started to ring. The locked doors cannot be opened until the carbon dioxide concentration has been reduced by use of the air conditioning apparatus (for details see AC Operating Manual Vol. 2, Sect. 8), and has been tested by switching on the special detector (see OC Vol. 1, Sect. 23, para. 6 for full instructions.)

Flashing lights with the word FIRE would have been preferable.

File conversion 10.4

File conversion is a system that involves problems of fact-finding, data capture, clerical procedure design, form design, and program specification. The way the job is done depends on the size and complexity of the files and the old and new recording media, which may involve manual systems of card index, folders, or binders; machine systems of punched cards, disk, or tape; master files or small specialized files of records; and centralized or scattered filing systems.

The result of the conversion job is the file the program specifies. Each field of records on the file will be identified with a source document. To go directly from the existing file to the new file is desirable, but usually impossible; a number of stages are often required. The ideal can be achieved only if the file to be converted is on a medium that is acceptable as computer input. Punched-card files can usually be converted directly, but even in this case, direct conversion is influenced by the quality of the existing cards, the code structure used on them, and whether they are truly acceptable input. For example, a punched-card file that has been maintained on 51-column stub cards cannot be used for direct input to a computer with a standard card reader. Similarly, existing magnetic-tape files must conform in size, layout, codes, labels, and so forth before a direct transfer can take place.

To transfer a file, the old file data is first clerically recorded on specially designed input documents and then transcribed to suitable media and verified. A tailor-made program is used to read the transcribed data; the file is then output in the format required by the user program.

One problem is that most source documents contain historical data that is irrelevant to the new computer file. Trained clerical help is usually needed to delete this information. Often experienced personnel must be released from operating departments to supervise procedures. It is sometimes possible to enlist outside data-preparation services for clerical transcription help. This approach may be wise when the technical content of clerical work is small or existing data-preparation resources are inadequate for a large conversion job.

Converting an active file such as an inventory file presents a special problem. It must be converted at a specific date so that data will be unavailable for the shortest possible time. An update procedure, involving additional clerical work, must be designed, and whether the file will be updated during the creation stage or on completion of the new file must be decided.

A final task that the analyst must organize is creation of the new files. The whole job may be done in one major effort or parts of the old records may be moved at intervals and then merged. The decision depends on the size of the task and whether time for the merge program is available. In all cases, a special program to establish the new file will probably be necessary. If the file is going to require updating, a special update program may also be needed. If so, the file may be created by treating all records as insertions, and using the standard file update program for the task.

Eventually, the analyst must check the accuracy of the new files. He must consider that they will be no more accurate than the original files, because errors may occur at each stage in file conversion. The analyst must ask: is the data in the source file correct? Are personnel available to check source records? What can they be checked against? Can we get control totals?

Rejections must be controlled to make sure all records are converted. Controls must be maintained until final cutover, although it is sometimes impossible to guarantee accuracy after cutover, because it is difficult to correlate data, and a manual file may be active during cutover.

The analyst must remember that it is essential to have some way of checking new files. The user departments must have confidence in them. They must not feel that the files were always reliable in their departments but, once in the computer department, are suddenly wrong.

Changeover   10.5

When the new system is fully tested and proved, changeover from the old system to the new occurs. This can be accomplished by direct changeover, parallel running, or pilot operation.

Direct changeover is the introduction of a completely new system without any reference to existing systems. This method is used only when there is insufficient similarity between the old and the new systems to make parallel or pilot runs or when the extra personnel required to supervise parallel runs are not available. A new system is usually introduced directly at a time when work is slack, so that affected personnel can become accustomed to the change without extreme conflicting demands on their time. To introduce a new system directly, the analyst must have previously established complete confidence in it.

Parallel running means processing current data by both old and new systems to cross-check results. The objectives of parallel running must be closely defined and a time limit set. If the method is to be an extension of the testing for the new system, it is useful only if the old and new outputs are strictly comparable, which is rarely the case. The difficulties of cross-checking must also be kept in mind. If a difference is found, the incorrect system must be identified. There is a natural desire to condemn the new system, but the error may be in the old.

Parallel running may also be used as a standby in case the new system breaks down. A firm time limit on the number of cycles for which the two systems will operate must be set. The cost and difficulties of running two systems and the possible effects on other computer work should be taken into account.

Pilot operation consists of two methods. One uses the new system with a previous period's data, so that results are known and can be checked. This test is easier to control than parallel running. The alternative method introduces the new system piecemeal, by phasing in the total volume of work gradually. This makes personnel transfer easier.

During this phase of the job the analyst should be especially aware of the coordination problems of the changeover. Communication lines should be kept clear throughout the system during changeover so that any errors that appear can be corrected. The analyst should remember that each change affects people, but that the system must be dynamic and capable of change. He must be aware of how many changes in the system and modifications in the program he is prepared to allow at each stage. He must also make sure that system maintenance methods are designed and working, so that it can be modified quickly and efficiently if necessary.

Exercises

1. In implementing a new system or application, what tasks must the systems analyst include in his plan?

2. What techniques are commonly used to plan and control the implementation process? What are the advantages of each?

3. Why is it usually necessary to use a computer to update a network?

4. Why will personnel requirements vary at different times in the implementation cycle? What alternatives are available to satisfy these requirements?

5. Discuss three types of reference manuals. Who are the intended readers of each type?

6. What problems are involved in converting manual files to machine-readable media?

7. What special problems are connected with converting a file that is constantly being updated?

8. What sort of controls are required to insure accuracy of converted files?

9. What are the respective advantages of the three types of changeover to a new system?

10. Assume you are a systems analyst assigned the task of preparing a bar chart to include the following pre-installation activities: new form orders, personnel training, site preparation, programming, and file conversion. Assign reasonable times to each activity and list them in logical sequence.

controls and security

Earlier chapters mentioned control features within the system. However, control is a broad topic, and some of its aspects are not logically related to a specific part of the system design process. This chapter emphasizes the importance of control and indicates some of the broader issues involved.

Data security consists primarily in the security of file information. Obviously without some assurance that the system's prime working data will not be corrupted or lost, the concept of an underlying control of the results of that system is meaningless. Hence, file security is the subject of a section of this chapter and is also discussed in the section dealing with audit controls.

11.1 Systems controls

Control can be defined as a means to insure that approved objectives are achieved. These objectives must be known and defined so that actual results can be compared with prediction.

Applying controls is a three-fold activity. The first step is to set the original measurements, or standards, for the planned objective. The achievement must then be measured against the standards, and if the measurement deviates from the plan by an unacceptable amount, corrective action must be taken. Thus, control is essential to achieve the objectives of the plan; but how are controls employed in a data-processing system?

First, the control system must be simple, so that it does not interrupt the usual flow of work. Therefore, it must be designed as an integral part of the system. Second, the controls must be essential and satisfy organizational requirements. The object of the control system is to maintain the system's objectives, and therefore response time is important. The ideal system is self-regulating and makes immediate correction, but this requires feedback.

Automatic feedback, or the introduction of corrections into the data-processing system automatically, is not always easy to achieve. Sometimes deviations from the plan require human interpretation and judgment before suitable system correction can be made. In this case, using the computer to make an automatic correction is not practical. However, in a forecasting system, for example, where the forecast is obtained by extrapolating past results by formula, the computer may modify the formula to account for errors in previous results. Thus a limited amount of self-regulation is performed. In designing a control system, a knowledge of system theory can be of assistance. An elementary introduction to this topic is included in Appendix B. Some elements of the control function have also been outlined in section 3.5.

There are three levels of control in data-processing systems. Management control is the system of controls management establishes to carry on the business of the company in an orderly manner, safeguard its assets and secure, so far as possible, the accuracy and reliability of records. Systems control is the subsystem that insures that the planned objectives of a system are being achieved. Procedure controls are inserted at various points in a system to prove procedures that depend on human and computer action are being carried out correctly. Obviously, both of the latter have to exist within the overall framework of management control.

Any controls in the system must detect the existence of errors, locate, and correct them. Examination of the data-processing system will show points where errors can occur. Plans about how to detect them and minimize their effect on the organization of the system may be made in advance to save time.

At the center of any data-processing system is the computer, where several kinds of errors can occur. For example, within the CPU there can be failures, power supply variations, power failure, interference on power input lines, and so forth. Also, many peripheral devices have moving parts working at very high speeds and very close tolerances, and these are subject to mechanical failure. Human errors, such as operators' and programmers' errors, can also occur, and the magnetic surfaces of file media can be subject to wear, damage by contact, and atmospheric problems.

The essential operation of any computer system can be considered to consist of data, processes, and results. Data can be subdivided into input and file data. A large part of the control activity must be centered in this area because this is the raw material upon which the computer operates. The GIGO principle applies to all processing. (GIGO is the acronym for "garbage in—garbage out.")

The object of the control system is to minimize errors, by self-regulation wherever possible, otherwise by continually measuring operations and issuing instructions for corrective action.

Electronic and mechanical failures are usually detected by the computer and software. For example, input failures on magnetic tape are often cleared because several attempts are made to read the same data. Preventive measures can be taken by special testing systems and maintenance. The effect of errors can also be minimized by anticipation—for instance, by the provision of frequent restart points within computer processing. Controls must also enable the data-processing system to *learn*. Learning helps avoid repeating mistakes. Any control system that the analyst designs must be able to answer the following:

1. Have we got all the required data?
2. Is the data translated from the source document correctly?
3. Is the data of the source document itself correct?
4. Has the data been processed all through the system?

Management controls   11.2

Senior management are not concerned with detailed control methods, such as those that make sure that old and new master file totals can be reconciled or that input documents have been expanded into the correct number of change records for the input edit program to update. They are concerned, however, with overall control questions associated with the systems that employ the computer. Their approach is that the computer is a black box, and the results of computer processing should enable them to see that policy limitations are under control. For example, if a policy decision has been made that inventory investment must not exceed a certain figure, the computer account system must provide an intelligible report that allows management to check that this limit has not been exceeded.

The controls imposed by management require information. One of the systems analyst's jobs is to determine the information that is needed. Information can either be essential, such as reports that generate immediate corrective action, or for interest, such as historical reports for future reference on which no immediate action is anticipated.

The analyst should pay more attention to essential information and to examining the corrective control procedures that exist. Essential reports must be given in time to allow action to be taken and should not be used primarily to fix blame for error.

Once the analyst clarifies the precise information requirements, he must decide whether *management by exception* techniques can be applied. Such techniques imply measuring deviations from previously established standards and reporting when a deviation exceeds previously set limits. When the system is functioning correctly, it will send only exception information to management. Because it is using information from other systems, management is assured that all policies of the organization are being carried out.

An example of this technique might be in the marketing area. Each product division with a company can be given a profit budget. The management information system is designed to report on any division that deviates from its budget by minus five percent or plus ten percent. If no reports appear, senior management can be assured that the overall company profit budget will not be more than five percent below the target figure. In this way, their time is saved. They do not have to read a lengthy report giving actual results against budget for each division, only to find at the end of the printout that all divisions are performing within their allowed tolerance limits. On the other hand, the appearance of a report is a significant event. Its presence shows immediately that at least one division has deviated from the plan and that some action is needed.

Even this simplified example shows how the interaction of the various subsystems within the computer can be used for management control. To achieve its profit budget, the division must sell the correct product mix at valid selling prices (data from invoicing system) related to the manufacturing costs (cost accounting system) using the correct number of administrative and selling staff (payroll data), each entailing a set amount of expenditure (data from expense accounting system). Thus the control report on profit budget automatically implies that the relevant parts of the other systems are in control. By the same token, however, the exception reporting system must be sophisticated enough to explain why a variation has occurred and in what area. Referring to the previous example, if sales division M were reported to be nine percent below its profit budget, little action other than starting an inquiry could be taken on this fact alone. But if the report showed that this deviation occurred because ten salesmen instead of eight had been employed, the sales manager concerned could be asked for an explanation. Similarly, if the variation were due to an increase in factory cost, the plant manager could be questioned. Obviously, to achieve this result, certain standards must be set for each subsystem and the management by exception method applied to them as well. The analyst must devise this set of interlinked controls at suitable levels. He must convince the many skeptics in management who often regard their status as measured in terms of the *number* of reports on their desks that the fewer reports they receive, the greater is their control over the overall situation.

Thus, management can use a data-processing system to control the organization by using it to plan, to set standards, to measure, and to report. The computer can collect information from various systems, carry out simple but numerous calculations, analyze, and simulate. These are familiar tasks, but a computer can provide results much faster than manual calculation.

The idea of using the computer to control an organization by employing it to test hypothetical policies and select the best one for introduction is gaining ground among management. These principles are usually called management science, and simulation techniques play a large part in many exercises of this type. The computer is often essential for the success of such ventures, because manual calculation is too slow and error-prone. Also, experiments with different policies can rarely be tried in the operating company, because of the confusion that would arise. Therefore, the management scientist constructs a model of the operating conditions that simulates as many features of the real-life situation as possible. The policies are then tried on the model and the results evaluated. In this way management has the opportunity to assess how practical a policy would be before it is introduced.

Model building, a specialized technique, relies heavily on accurate facts about the existing situation, many of which are usually obtained during routine systems analysis investigations. There are two kinds of models: those that use past history for projection into the future, such as inventory control/forecasting, factor analysis, and distribution models; and those that state the existing structure as a set of mathematical relationships that can be tested to evaluate alternatives, such as vehicle routing and queuing theory models. A number of excellent books have been written on these topics. The analyst is referred to these for more information.

11.3   Audit controls

The owners of an organization are also interested in control. They appoint auditors to examine and report on the efficiency of the whole system of controls. (In computer systems, however, they will rarely examine records that are on tape or disk or check procedures in the form of program instructions.)

An audit trail should be established in all computer-based systems so that the details underlying summarized accounting data, such as control and reconciliation totals, may be obtained. This trail must be included at the design stage. Management will also be interested in this data in the event of outside queries, for example, when invoices and statements disagree with customers' calculations.

An audit trail must extend through the system from data collection to result distribution, and can often use many of the controls already established. The problem of files changing frequently and old versions being destroyed, thus causing a break in the trail, can be overcome by printing out changes and filing the printouts so that a complete external history is available. Another method is to keep additional copies of files for audit purposes, as discussed in section 11.5 on file security.

Procedures usually exist in program form. Unless the auditor is expert, he will not understand program technicalities. For this reason, he will be interested in the test system, the test data, the results achieved, and the security methods that insure the approved version of the program is being used. The processing log should show program identity, including the version number, and should also include details of mechanical failures and of action taken.

Some auditors have secure test decks that give known results with a particular set of programs. When performing an audit, they run the test deck with the program and check the results against the original set. In this way, they can make sure that unauthorized program modifications have not been made. Although this theory is sound, it is not completely successful in practice. In day-to-day working, many minor program modifications have to be incorporated. These could invalidate the auditor's test data, because his visits tend to be infrequent. Consequently, many auditors satisfy themselves that security procedures for the introduction of program modifications are adequate and that authority for them is given at a suitable level.

Audit problems are more acute in real-time systems where no history is kept and control is immediate. Similar problems are involved in recording the actions of a self-regulating system with self-correcting ability. However, the auditor will usually accept a reasonable control system. Thus, the analyst must provide management with suitable audit controls. A number of specialized texts on this aspect of the subject are listed in Appendix C.

Controls of tests   11.4

Section 7.1 referred to the tests programmers perform to validate the complete system and to test their data requirements. Close control must be maintained during this phase of the work so that tests are completed within a reasonable time and changeover can begin. There are two kinds of tests to be considered: link tests and operational tests.

Link tests are concerned with testing the programs to see that they behave satisfactorily when given data prepared by other programs in the system, as distinct from data prepared by the writer. Pairs linking is the first stage. All programs that feed data into a given program are identified. The program is then combined with each feeding program, in turn, to form a series of pairs. Each pair is tested on the machine. The test data for these tests should be all or part of the program test data constructed for the feeding program. Its output is fed into the program being linked. Errors in either of the programs may be detected at this stage.

The first technique for detecting errors during link tests should be to discover the essential difference between the data being fed from the feeding program and the original program test data that represented it. There are four broad categories of difference. Quantity of data may be different, or file layouts of the same files may differ between programs. Even if file layouts are the same, the writers of two programs may understand the functions of certain fields differently. Also, untested conditions may exist in the data.

After the essential difference is identified, it should be possible to recognize the part of the programs affected, and hence the cause.

System linking follows pairs linking. In this stage, the test data prepared by the systems analyst is punched and submitted to the system as a whole. Before this can be done, suitable files must be prepared by using the system's file addition facilities. Again, errors should be detected by recognizing the essential difference between the data now being produced and test data.

Even though the set of programs has been linked and tested with system test data, it would be imprudent to use the new system in place of the old system until the set is tested for a considerable time under realistic conditions. A large quantity of current or past transactions should therefore be fed into the system.

These operational tests check not only the programs but also the data-preparation and all other related manual systems. Errors will be found in the programs, in the proposed system, and very possibly, in the system it replaces. Many employees, including systems analysts, will be involved in this operation, which is quite expensive, since three operations are performed simultaneously: both the new and the old systems are run, and the results of the two are compared.

This period should be short. Everything possible to insure that the parts are in good shape before operational tests begin should be done during individual program and link tests.

Program errors should be detected and corrected by the programmers. The essential-difference technique described previously is usually helpful here, but some manual changes may be necessary, since different programmers are often involved at this stage.

A Program Test Log (see Figure 11-A) should be kept for each program. On it, each test, assembly, or link test is numbered and the date of submitting the program is shown. The type of errors that occur is recorded for each test.

The time recorded on the copy of the operator's log returned with the completed test is entered under "Elapsed Time." The time agreed to be charged for the test after investigation of the results is entered under "Effective Time."

PROGRAM TEST LOG

PROGRAM:

Assembly No.	Test No.	Link Test No.	Mis-operation	Computer Fault	Data Preparation	Data Error	Test Data Error	Assemblies		Tests		Link Tests		Notes
								Total Standard Time		Total Standard Time		Total Standard Time		
								Elapsed Time	Effective Time	Elapsed Time	Effective Time	Elapsed Time	Effective Time	

Fig. 11-A. Program test log

The program name and number are entered at the top of the sheet with the standard time for tests, assemblies, and compilations.

In addition, a system test as shown in Figure 11-B should be maintained by the lead programmer, or a senior programmer in charge of a set of programs, in order to:

1. Insure that the number of tests shown by each programmer in his program test log is correct;
2. Determine the accumulated computer time to be charged for tests;
3. Measure the delay between submitting a test to the computer and its completion and return;
4. Record the date the test actually took place; and
5. Establish an overall daily progress position for all programs.

File Security   11.5

Once a system requires a file, the problem of file security arises. Since the information recorded on the file is vital to the efficient working of the total system, it must not be destroyed or changed.

SYSTEM TEST LOG

SYSTEM:

Date Sent	Program	Date Performed	Date Returned	Time Elapsed		Time in Query	Adjustments		Remarks
				This Test	Total		This Test	Total	

Fig. 11-B. System test log

Security is exercised at two levels. Physical security consists of making sure that no damage is done to the files while they are awaiting processing. Operational security insures that data is not lost during processing.

Physical security is usually the operations supervisor's responsibility. Part of his function is to make sure that files are stored properly and protected from fire. If magnetic media are employed, he is responsible for keeping them free from dust, in a suitable atmosphere without excess temperature or humidity, and away from stray magnetic fields. He must also make certain that correct files are issued to the operating staff and that proper records of the movement of all files are kept. All tape reels must be suitably labeled to give adequate identification of their contents. All files should have expiration dates, and each should display the date on its file label as an added security check. The supervisor must also instruct operators in the correct methods of handling media to avoid such damage as creasing magnetic tapes or tearing punched cards.

Operational security is a joint effort of systems analysts, programmers, and equipment manufacturers. Manufacturers often provide hardware devices to assist operators, such as the file-protect ring to prevent data being written on a file.

Hardware can also perform parity checks on individual characters after they have been processed. The parity checks are used to detect invalid bit patterns. When the sum of the bits that make up the character *plus* the parity bit must total an even number, the system is called even parity checking. A similar method—to total an odd number—is known as odd parity checking. A parity failure implies that there has been some hardware malfunction causing the incorrect transfer of data. Processing stops and an error signal is displayed. Before this happens, however, the program usually loops back to an earlier point, recovers the original data and attempts the processing again. Techniques of this type are called hardware protection. They might consist of performing numerical reconciliations after a certain number of records have been processed, and abandoning the run if the checks fail. It is usual to dump the contents of internal storage at the time of the failure, so that the program may be restarted at the same point when the cause of the failure is discovered. In some circumstances, the analyst specifies logical points in the processing where meaningful checks can be carried out. Some care must be taken not to overload the program with checks, because restart situations can be expensive even if the files are secure.

When using direct-access files, possible restart action after a processing error is detected must be determined. If a file is updated by copying it in a new area (see section 5.3) the original file is still available after updating. With processing by the overlay method, however, more care is needed. The contents of the old file are overwritten and unavailable if the new file proves to be incorrect. This problem can be overcome by copying the original record into a temporary storage area until the processing has been checked and proved correct. If there is an error, the original file is still available, and processing can be repeated. If no error exists, the temporary storage area can be used to copy the next set of original records from the file while they are being updated.

Systems analysts must often devise suitable file-control totals that can provide useful safety checks. Reconciling old master file totals plus transaction totals against new master file totals can highlight errors before the situation is irrecoverable. There are two main types of control totals for files: quantity, or value, totals are meaningful and may be valuable in actual processing; hash, or nonsense, totals have no meaning for processing, other than for reconciliation.

An example of the use of both types might be found in a cash receipts portion of an accounts receivable application in which customers send their monthly remittance to the computer center, where account numbers and amounts are recorded for input. Each batch of input contains a control slip containing two totals, which are also punched. One total is a hash total of relevant customer account numbers and the other is a value total of amounts received. During the updating of the accounts receivable file, the computer stores the total of the account numbers of active records, the total of amounts recorded on those records, and the current value of all outstanding balances. When the updating run is finished, three methods of reconciliation are possible.

1. The input hash total could be compared with the total of account numbers processed, to insure that all relevant accounts have been updated.
2. The total of the amounts recorded could be compared with the input value total to confirm that all amounts have been written to the file correctly.
3. The new total outstanding balance could be calculated from the balance on the old master file minus the sum in (2) above, to doublecheck that recording is accurate.

Another file security technique applicable primarily to magnetic tape files is the grandfather-father-son method. With this scheme, three versions of a file are available at any one time. File 3 (son) is the current file which was created from file 2 (father) which, in turn, was created from file 1 (grandfather), as shown in Figure 11-C.

The advantage of this technique is that recovery is always possible. For example, if the new data on file 3 contained errors, the job could be repeated using file 2 again with transaction data. If both file 2 and file 3 were damaged during processing, file 1 is still available to create file 2, which can then create a new file 3. If file 3 contains no errors, it is used as the father file in the next processing cycle to create a new son file, which will be physically output on the old file 1.

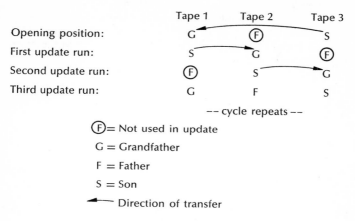

Fig. 11-C. Generations of files for security

Direct-access devices theoretically require more security features and routines than magnetic tapes since many of the safety measures used with tapes are absent. For example, several files may be kept on one disk, programs may be kept with other files, updating is often by overlay, and in most cases no grandfather-father-son files are kept. In practice, direct-access systems may be just as reliable as magnetic tape systems because hardware checks and software routines are provided by the manufacturer. However, because of these inherent processing differences, both systems analysts and programmers should be particularly careful to insert additional checks into all programs that use direct-access facilities.

In addition to the labels on the external containers of storage media, file labels are incorporated as part of the recorded data itself. A label acts as the first record on a particular file, and contains information allowing the computer to check that the correct file has been set up. A date on the label is often compared with the current processing date, or sequential run numbers are used. Expiration dates also appear on the label.

On magnetic-tape files, the label is usually the first physical record on the tape. With direct-access files, particuarly disks, all labels for all files are in one place. In addition to the usual checking information, the label indicates the location on the disk of the first data record of the file. The software control features, which are used to prevent unauthorized reading or writing in the same way as the hardware devices on magnetic tapes, are also in the label of a direct-access file. Figure 11-D illustrates typical file labels for both tape and disk, showing the kind of information usually found on them.

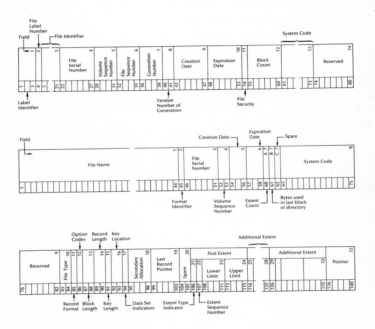

Fig. 11-D. Magnetic tape and disk labels

Dumping file contents periodically is another technique used to improve security. The contents are copied onto another physical storage medium that can be housed separately from the file used in processing. The method is applied particularly to direct-access files that are updated by overlay, and to transaction files that may be needed if reconstruction of the master file becomes necessary.

Dumping can be a time-consuming routine. Magnetic tape is usually a good receiving medium for a dumped file, particularly if dumping can be accomplished as a by-product when the file is read from disk.

Duplication of files is a security technique that can be helpful when working with large files. The method is to make changes to the master file and copy the new version on a second device at the same time. For example, the old record can be overlaid on disk, and the updated disk record can be copied on magnetic tape.

In rare cases, complete duplication of files may exist when two exact copies of each file are maintained. This could be necessary for extreme accuracy in an interrogation system.

Some physical security measures should be taken to insure that the initiating documents are not lost or damaged. If questions arise after the file has been updated, one should be able to check that the original instructions were followed. Sequentially numbered documents that are retained on file at the computer installation, or in the originating department, often provide the required control.

Exercises

1. In applying controls, what three activities are required?

2. Automatic correction of errors is ideal, but not always feasible; why?

3. Explain the differences in the three levels of control in data-processing systems.

4. What is meant by *management by exception?* Give an example, from your own experience, if possible.

5. Discuss model building as a management science technique. Why is it particularly adaptable to a computer?

6. Comment on the following statement: "As long as there are no complaints about the accuracy of results produced by a computer, the internal auditor will not be concerned with how these results are produced."

7. Explain the difference between a link test and a systems test. If each program has been checked out using programmer-supplied test data, why should it be necessary to perform link or systems tests?

8. Compare the precautions required for operational security of files stored on magnetic tape versus direct-access media.

9. Receiving and Shipping Authorization slips are used to record the receipt and withdrawal of parts from inventory. If cards were to be keypunched from these slips for further processing, what data fields could be accumulated as control totals?

10. What types of information will file header and trailer labels contain? As long as the computer operator can mount the correct tape reel or disk pack by referring to its external label, why should it be necessary to have header and trailer labels with the file itself?

12

hardware

A full discussion of computer hardware would require a book in its own right. Each manufacturer has ideas on the best design characteristics, and computers made by different companies are not strictly comparable. New machines and peripheral devices appear so frequently that it is impossible to possess a thorough knowledge of all current hardware.

The trainee systems analyst need only concern himself with the basic details of central processing units and peripheral equipment. He is unlikely to become involved with their finer design points. He need only be able to follow any discussion of the topic. For this reason, the first section of this chapter deals with hardware items in broad outline only. Manufacturers will supply detailed technical literature for the analyst who requires more information on the capabilities of a particular model.

However, the analyst is likely to become more closely associated with punched-card, data-transmission, and paper-handling equipment, and these types of hardware are discussed in greater detail.

12.1   CPUs and peripheral equipment

A computer is sometimes defined as "any device capable of accepting data automatically, applying a sequence of processes to the data, and supplying the results of these processes." This broad statement is divided into more restricted definitions such as that of a stored-program computer, which is defined as "a computer capable of storage of all or some of its instructions, so that the stored instructions may be altered within the computer," and that of a digital computer, which is defined as "a computer in which data is represented in digital form." Throughout this book, the term *computer* has been used to mean *stored-program digital computer*. Analog computers, in which variables are represented by physical quantities, and hybrid computers, which are a combination of digital and analog machines, are excluded.

A computer consists of five basic elements, as shown in Figure 12-A. The input and output sections of the diagram

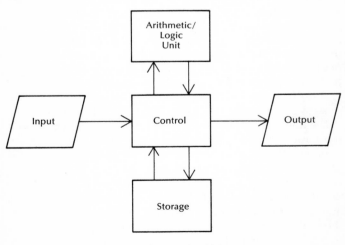

Central Processing Unit

Fig. 12-A. Basic elements of a computer

represent a combination of devices used to read data into storage or write data from storage. Some of them are:

1. Punched-card readers and punches
2. Paper-tape readers and punches
3. Magnetic ink character readers
4. Optical mark and character readers
5. Line printers
6. Character printers
7. Computer-output microfilm
8. Cathode ray tube units
9. Direct entry consoles and recorders
10. Graph plotters
11. Audio response units

The various magnetic media are also used extensively for input and output during file-handling activities. They are considered below in the description of storage devices, since it is unusual to use them for prime input or output.

Before considering the storage element in more detail, the physical means of inputting and outputting data in the form of electrical impulses should be considered. Each device that can be attached to a CPU is governed by a control unit provided with a standard peripheral interface so that different input/output devices can be connected to the same CPU. The connection between the input/output device and the standard interface is called a *trunk*. From the interface, the data signals must be transmitted to the control section within the central processor. This final path is known as a *channel* (see Figure 12-B).

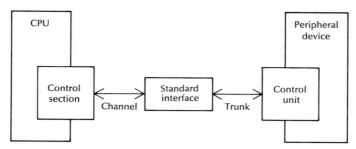

Fig. 12-B. Physical means of data transfer

The storage element in the computer consists of internal and external parts. The internal storage of the machine is usually called primary, or main, storage. It is this part that is referred to when sizes of machines are quoted. At the present time, magnetic cores are used extensively for main memory, but plated wire, magnetic rods, and especially thin film are being used increasingly. It is necessary to know the way in which any particular machine allocates its main storage if comparisons are to be made. Although all information is stored as binary digits or *bits,* the smallest addressable location in main storage differs from one machine to another. These storage locations may be—

1. Bytes—one 8-bit pattern of information, used to store either one alphabetic or special character or two numerals;
2. Characters—one numeral, one alphabetic character, or one special character; or
3. Words—an ordered set of bits which may be fixed or variable in length, depending on the machine. Usually expressed as a number of bits.

Access and cycling should also be considered when discussing core storage. The access time for a machine is the time interval between initiating a transfer of data to or from storage and the completion of that transfer. The cycle time, on the other hand, refers to the delay that must occur between one access and the next. The amount of information that can be accessed in one cycle depends on the type of machine involved, and is expressed in bytes, characters, or words. These factors have an important bearing upon a program's overall processing times, even though they are measured in *micro-* (millionths of) and *nano-* (billionths of) *seconds.* Access and cycling times depend on the physical medium used for storage. Sometimes main storage is designed on a modular principle, so that its size can be increased if the system expands. When this is impossible, additional peripheral devices can be added and treated as extensions of main storage (as auxiliary storage). Magnetic drums, disks, and tapes can be used for this purpose.

These and other magnetic media are more commonly used for external storage. When used in this way, their characteristics are relevant to file media, file handling, and file organization, which have been discussed in chapter 5.

The control element of the computer consists of both hardware and the operating system. The latter will be discussed more fully in the next chapter. It must be present in core storage for control during the execution of any program, and it reduces the effective size of the core storage area.

The arithmetic/logic unit in the central processor works in a very elementary fashion. It is capable of addition or subtraction of two patterns of electronic pulses only, and performs multiplication by repeated addition and division by repeated subtraction. Because of the basic simplicity of these units, the programmer must describe every job in its most fundamental terms. In consequence, the systems analyst must provide the programmer with a detailed specification to show how the computer system is to function. Breaking down a job into such fine detail involves careful analysis of every requirement. No logical errors must be present in the finished specification.

Although basically a simple device, the arithmetic/logic unit can work at extremely high speeds. In comparison, the input, output, and external storage devices are relatively slow; getting data into and out of the system is time-consuming. Table 12-C compares speeds of the various devices.

To overcome this incompatibility in speeds, alternative methods of processing have been devised. The basic rules of

TABLE 12-C

APPROXIMATE PERIPHERAL DATA-TRANSFER SPEEDS

Peripheral device	Data-transfer speed (characters per second)
Console typewriter	10–20
Line printers	750–2,420
Paper tape reader	1,000–2,000
Magnetic tape—slow	30,000
Magnetic tape—medium	60,000
Magnetic tape—fast	340,000
Replaceable magnetic disk	156,000–312,000
Fixed magnetic disk[a]	156,000
Magnetic drums[a]	1,200,000

([a]Mostly used as auxiliary storage)

sequential operation in which one task follows the next are bypassed. The methods include—

1. Simultaneity—in which different elements can be working at the same time. For example, punched cards can be simultaneously read for Job A, calculations can be performed for Job B, and printing can be done for Job C.
2. Multiprogramming—in which the CPU is executing more than one program concurrently, by interrupting one in favor of another at logical points, such as when the first program must pause for the input or output of data (see also chapter 13).
3. Multiprocessing—in which more than one central processor is employed in the system.

All of these approaches increase system complexity. The whole operation must work at the speed of the fastest unit if any advantage is to be gained. Possible delays due to human intervention must be reduced to a minimum. This may impose very strict discipline on sections of the organization that use the data-processing system.

12.2   Punched-card equipment

As mentioned in section 3.2, punched cards and punched paper tape are common input media to computer systems. Most computer users formerly used conventional punched-card systems. The analyst will often be in close contact with such systems and should be aware of the broad principles governing the equipment currently being used in his organization. This section will provide a basic introduction to the equipment; manufacturer's manuals will give more specific information where required.

The main functions performed by punched-card equipment are punching, verifying, duplicating (of code patterns), interpreting, sorting, collating, calculating, and tabulating.

Card punching is the basic means by which source data is converted into punched cards. The operator reads information from a source document, and by depressing keys, converts the information into holes in the card. There are two types of card punch, hand and automatic. A simple hand punch consists of a carriage carrying a single card and a set of punches operated by a group of keys. Information is punched in one column of the card at a time. Movement of the carriage is controlled by the operator when more than one column has to be bypassed. Hand punches are suitable only when small quantities of cards require punching.

With an automatic punch, data is also entered on a type-writer-like keyboard, but the punch is power driven. This reduces the operator's physical effort. The cards are automatically positioned and fed through the machine. Certain punches can print the punched information on the top of the card directly above the punched holes. Each column may be punched separately, or the automatic punch may punch the complete card in one operation, and enable corrections to be made during the punching of the card. In general, automatic punches are capable of duplicating common information, such as the date, into every card of a particular batch.

Verification is simply a means of checking the accuracy of the keypunching operation. The punching operation is repeated using a verifying machine, which is very similar to a punch, but, instead of punching, it compares keyed information with that already punched in the card. The verifier signals any discrepancy between the information on the card and the keys depressed by the verifier operator. Corrections are made by keypunching correct cards and discarding the error cards. To distinguish verified cards, the verifying machine punches a small notch on the right side of each correct card after verification.

If the information to be punched is primarily numeric, a check-digit verifier feature can be installed on a card punch. This feature will check that control information, such as account number, has been correctly keyed by the operator. In effect, the verifier is a small electronic calculator.

Duplicating is the automatic punching of repetitive information from a master card into a batch of detail cards. This is usually done as part of the punching function, and most card punches can perform it. The main advantage is that the operator keys common information, such as date, account number, and so forth, only once at the start of a batch. This reduces the chance of error and speeds up the operation.

Gangpunching is an extension of the duplicating function

in which information from a master card is automatically punched into all following detail cards. The main difference is that a master card is read and detail cards are punched: that is, the equipment is a combined reader/punch, not simply a punch, as in the case of duplicating.

Reproducing copies information from one card to another by a reproducing punch, which is basically a reader/punch. Mark sensing is the automatic punching of cards from marks made on the card by a special pencil. In this case, the card has probably been used as a source document in the system.

Interpreting is the translation of punched information into printed information on the card. It can be done by an interpreting machine or a keypunch equipped with a printing mechanism. If interpreting is done as the card is punched on a keypunch, each printed character will appear directly above the column in which it is punched. If a separate interpreting machine is used, each character may be printed anywhere across the top of the card, regardless of where it is punched.

End printing is very similar to interpreting, but the information is printed in large, bold type along the end of the card instead of on the top edge. When cards are used for prepunched files, and are stored on end, this form of interpreting is useful for speedy reference and selection.

The grouping of cards into numerical or alphabetical sequence is known as sorting. Sorting machines are equipped with several pockets to enable groups to be isolated, or to select individual cards from a deck. The machine is set manually to perform specific selections, usually examining one column at a time.

A collating machine combines groups of cards or matches cards from different files. Essentially, the cards are extracted from two hoppers on the machine and information in selected columns is compared. In the case of merging, two card files in the same sequence are fed through the machine and combined to form one complete sequential file. For the matching operation, the machine checks certain information from two sets of cards and rejects unmatched cards.

Calculating machines are capable of reading information from cards, performing certain basic arithmetic functions, including multiplication and division, and then punching out the results of the calculations on either the original card or a trailer card.

Tabulating machines read cards and print information from them on continuous forms or single sheets. It is possible to perform simple additions or subtractions with the tabulator so that accumulated totals may be printed. This process is often used in conventional punched-card invoicing applications to arrive at net prices.

In practical situations, a punched-card system may include preparing cards, pulling them, interpreting them, sorting, selecting, and merging them, calculating, and tabulating.

Card preparation is concerned with the transfer of data from a source document. The data may be in the form of transactions or file information. For example, orders received have to be punched into cards. Customer names and addresses must be placed in a name and address file that is updated as required. Any other fixed data can be kept in the form of a card file and used for reference. Any information that is punched in cards must be verified to minimize error.

When card files of standing data are kept, the card-pulling operation becomes necessary. The cards in the pulling file must be interpreted so that certain cross-reference details may be visually checked. The pulling file must be up-to-date so that errors are not introduced from this source.

In some systems, cards are produced at one time and re-enter the system later. For example, in an inventory application in which a finished assembly is analyzed into its component parts with a master reference deck, the set of detail cards could be sent to inventory for issue. As the parts are released, the parts clerk might enter the date and quantity on the original card and return it to the punched card center for tabulating on a daily issue list. For applications of this type, the card must be interpreted.

In any system, certain information must be processed in a definite sequence. For instance, certain management reports may include a print of sales by area or product. Thus, at certain parts of the system, data must be arranged into a particular sequence in readiness for the next operation. This is known as sorting and is a considerable part of any punched-card system.

Another operation concerned with arranging cards is selecting. This process isolates particular cards from a deck (for example, master cards that must be extracted after use and returned to the pulling files). Selection can be performed by either a sorter or a collator.

Merging is the opposite of selecting. It involves bringing together two decks of cards, essentially in the same sequence, and combining them to form one deck. For example, cards may need to be extracted from a name and address file and combined with the corresponding customer orders from a transaction file. A collator can accomplish this, and usually it can also reject unused or unmatched cards.

Though certain basic information may enter a system as new data (for example, the quantity ordered of a particular product), other information is obtained from tub files or permanent card files in the system. Thus, discount rates for each customer and unit prices for each product may typically be kept as fixed data in the system. To invoice the customer, the net invoice amount must be determined. This requires several calculations, namely, multiplying the unit price by quantity and subtracting any discount to arrive at an invoice total. For simple additions and subtractions, a tabulator can be used, but for multiplication and division an automatic calculator is required. The results of the calculations can be printed directly or punched into cards for further processing in the system.

In general, the tabulator can be considered the system's printing device. It produces any printed output the system requires, such as batch totals, invoices, sales statistics, statements, and so forth.

Punched-card equipment may affect the computer environment as an input or output medium or as a complementary system. Punched-card equipment in a computer installation can perform much of the system's functions, such as reproducing or collating, as offline or auxiliary equipment. Card readers and punches that are online to the computer operate much faster than their counterparts in the conventional system. The conventional equipment, such as sorters, collators, calculators, and tabulators, is not usually required for computer systems unless the punched-card system is running in parallel temporarily, as, for example, during conversion to a computer system. One exception is that some computer installations that use card input retain a conventional sorter to ease their CPU's sorting load.

Data transmission   12.3

Data transmission is one of the fastest developing techniques in data processing. As knowledge of the use of computer techniques grows, so does computer usage. Furthermore, the growth of this knowledge frequently calls for faster and bigger computers. This manifests itself in two ways. Often an organization needs to collect and process information from different physical locations more quickly than possible by traditional methods of communication such as messenger or postal service. Also, an organization or individual may sometimes need access to a large computer for processing its own programs and data quickly.

The choice of data-transmission equipment raises a number of questions about the system itself. How much information is to be gathered? Is it to be gathered at fixed times, at any time, in batches or by individual record? How many locations will send information? How far are the locations from the computer—in the same building, in the same area, or many miles away? Will data flow in only one direction (simplex)? Is transmission from the computer to the terminal required? At the same time (duplex)? Not at the same time (half-duplex)? Is the terminal to be online to the computer? Is it preferable to produce the data in some form of storage medium (tape/cards) at the computer center before process time? Must the remote user have the ability to develop his own programs at his console? Can the cost of data transmission be justified? Can the cost be reduced by limiting the number of terminals?

Data-transmission lines may be leased from AT & T, Western Union, TWX, or IBM. Teletype channels transmit data at a speed of between five and twenty characters per second and are suitable for handling small or moderate volumes of manually-keyed data. On telephone lines, up to 300 characters per second can be transmitted, depending on whether a public or a private line is used. If there is a large volume of data, it may be convenient for a company to lease private lines for transmitting both voice and data. Broad-band channels of high-frequency electrical signals can transmit 100,000 characters per second, but this method is quite costly and is suitable only when a large volume of data must be transmitted fairly steadily.

A variety of terminal and control equipment is available from major computer manufacturers. Some of the devices are suitable for many purposes, while others are intended for use in specific application areas.

The following lists give an indication of the wide scope of equipment. Manufacturers will supply detailed specifications if the analyst wishes to consider data transmission for his system.

1. Possible methods of input at the terminals are:

Punched paper tape
Punched cards
Magnetic tape
Plastic badge
Keyboard
Telephone dial
Light pen
Encoders
Lever selection
Edge-punched cards

2. Possible output at the data-processing center can be:

Punched paper tape
Punched cards
Magnetic tape
Hard copy
Magnetic disk
Visual display

12.4   Paper-handling equipment

One disturbing feature of commercial data-processing installations is the tendency to output very large quantities of paper. Reports, statements, invoices, purchase orders, production plans, analyses, and so forth appear from high-speed printers at an alarming rate. The analyst's responsibility does not end after he has designed an acceptable printout. He is also responsible for ascertaining the numbers of copies required, distributing them, and designing a system to provide the information in the physical form required with sprocket holes trimmed, continuous sheets separated, and interleaved carbons removed.

Fortunately, a wide variety of paper-handling equipment is available. Decollators are machines that will remove the interleaved one-time carbons from multipart continuous sets and separate the individual parts. These parts are refolded into individual trays, so that the original multipart set becomes a number of one-part sets in continuous form. The machines usually trim off the sprocket holes on both sides of the form, but this operation can be performed on a trimmer if necessary. Variations of the standard decollator can remove one part from a multipart set while leaving the balance of the set intact.

Once the carbon has been extracted, the individual forms must be separated from their continuous web by a machine called a burster. It tears the forms apart at their perforations and stacks them ready for distribution. One interesting adaptation of the device is the burster equipped with a signature plate that enables the separated forms to be signed before stacking.

Most of the machines described above are available as freestanding units to perform the operations required to convert the printer output into acceptable documents for distribution. However, many of the individual units can be linked, in which case they are more properly described as combined machines. This arrangement allows the user to build up his paper-handling equipment to the exact specification required for his installation. Nearly all the machinery can be equipped with optional extras for unusual jobs. The manufacturer's sales literature should be examined carefully before selecting a particular unit.

When trimmed output has been obtained, the next problem is how to distribute it. Once more, specialized equipment is available to handle the task. Paper-folding machines will fold individual forms to the size required for standard envelopes. Advanced models can collate and fold two or more forms together in the same way. To avoid the need for manual envelope insertion, some types of paper-folding equipment can perform this job automatically. One feed hopper is loaded with the output forms, a second hopper takes the unfolded envelopes, and the machine folds, inserts, seals, and delivers the envelopes ready for mailing. Even stamping can be automated by using a postage meter.

At times, particularly with internal reports and documents, the analyst's problem is not one of distributing multi-output, but of producing a large number of copies of one document. There are many different kinds of copying equipment, but there are basically only two ways of tackling the problem. The computer output can either be produced on a special form that can be used as a master for subsequent copies, or, by printing a single copy on standard paper, copies can be made on conventional office copying equipment.

The use of special forms includes output printed on hectographic carbon for spirit duplicating, or on paper masters for offset duplicating. Each method has its advantages, and, depending on the situation, each is an efficient means of producing multiple copies.

If single copies are produced on the printer, the choice of copying processes is the same as for producing copies from any original office document. One limitation on many office copiers, however, is the size of the original document. They are often designed for copying legal-size papers as a maximum, and computer printout sheets are usually larger than normal office paper sizes.

The final consideration of paper handling is how to store the printout until needed for reference. Specially constructed binders provide one solution; they have elastic cords threaded through the sprocket holes in the print margins and clamped under metal clips, allowing the pages to be opened, so that none of the print is obscured in the paper folds. There are also patented filing methods on the market.

If the bulk of the printout is particularly large, some computers may be specially equipped to output data directly on microfilm. Cameras that will accept continuous forms and copy their content on standard film are also available. Microfilming reduces storage space, but care must be taken in the design of the indexing method on the film, so that retrieval of the stored data is simple. Normal office microfilm equipment is suitable for copying printouts after they have been separated and converted to individual forms, subject to paper size limitations referred to above in connection with all copying processes.

By careful appraisal and recognition of the functions of various equipment, the analyst can propose an integrated system capable of handling any output from the computer. The services of manufacturers' specialist representatives are available to him in cases of doubt.

Exercises

1. What is the function of each of the five basic elements of a computer system?

2. What is the difference between byte, character, and word storage?

3. Input/output devices are relatively slow, compared to the CPU. What processing methods have been devised to overcome this limitation?

4. Why do magnetic storage media, as a group, have a much faster character transfer rate than punched cards or paper tape?

5. What functions does punched-card equipment perform? Give an example of an application involving each function.

6. What factors must the systems analyst consider in deciding whether to include data-transmission equipment in his systems design?

7. Discuss the functions of three types of paper-handling equipment.

8. What problem arises when computer output forms are to be duplicated on conventional office copying equipment?

9. Contrast the advantages of outputting data on microfilm rather than on a line printer.

13

software

Many sets of coded instructions or programs have been designed to fulfill a wide range of common requirements and make the best use of computer hardware. Such sets of instructions are called *software*. Software is normally supplied by manufacturers, but parts of it may be developed by the user to meet specific requirements. Regardless of its origin, it performs two main functions. First, it enables the internal control functions of the hardware to be carried out efficiently at the speed of the fastest equipment—the CPU. Second, it makes programming easier and this, in turn, reduces the delays before a user's machine can become fully operational and productive.

There is a considerable lack of standardization in software. Some of it is peculiar to one machine only. However, certain general principles have been followed in its development, and this chapter is concerned with them. It may be regarded as a survey of the subject, rather than a detailed specification.

Because software covers such a wide range of requirements, it is advisable to subdivide it into a few main types. The sections that follow discuss each of these in turn.

Operating systems 13.1

As noted in the previous chapter, online computer hardware is a complex arrangement of individual units working together to manipulate data at very high speeds. To control the necessary actions efficiently, one group of software is concerned with the hardware organization. This group controls the input and output devices and internal data transfers, and insures that required peripheral equipment is available for use at correct times.

This becomes of vital importance when multiprogramming is being used in the configuration (that is, when several running programs are in the computer at the same time). For example, one program may be sorting data, a second performing calculations, and a third printing results of an earlier program from magnetic tape. All these programs are using the same CPU but time is shared according to relative priorities. The decision-making required to monitor a combination of programs successfully is controlled by an executive or supervisor program. The basis of the arrangement for time sharing is that one program is allowed to run to a logical point when it may be interrupted. Then, all programs are examined and that with the highest priority is executed until it reaches an interrupt point. The cycle is repeated until all programs have been completed. For obvious reasons, this method of control is usually called an interrupt system. It insures optimum use of the CPU.

Another class of software within operating systems is devoted to detecting errors caused by machine malfunctions. For example, if a parity failure occurs when data is read after transfer from one storage location to another, the software will cause the operation to be repeated in an attempt to clear the error. If it remains, the program will be stopped and a suitable error signal transmitted to the operator. Some software of this type has facilities for automatically printing out the entire contents of internal storage in machine code when a failure occurs, so that the programmer can investigate. This operation is commonly referred to as dumping.

Dumping is also of considerable assistance to the programmer in the second form of error detection: detecting errors in the logic of a program. When faulty logic causes recognizable error conditions within the machine, the software takes over and reports on the situation. Examples of this type of error are those caused by:

1. Register overflow as a result of a calculation step;
2. Incorrect modification of storage locations, which causes an attempt to transfer data outside the limits of a previously defined area; or
3. Attempting to use peripheral devices that are not attached or are assigned to another task.

This software is closely allied to that designed for tracing the operation of a program. These diagnostic routines will print out the results of each instruction as it is executed with a suitable commentary that the programmer can understand.

Another part of the operating system is devoted to communication with the operator. Most commonly such a man-machine relationship is achieved by a console typewriter, which prints out a log commentary of significant events plus requests for commands. This log includes details of the programs being run, the peripheral devices in use, file identifiers, and running times. When an error occurs, or a programmed interrupt point is reached, a suitable message is typed on the log. Using stylized forms of address, the operator instructs the machine of the next action required by keying a combination of letters and numerals on the keyboard of the typewriter. In some advanced applications, these questions and answers appear as English sentences, thus easing the operator's task.

Operational efficiency is very important, since considerable time can be wasted in changing from one job to another. Job-control software attempts to minimize this wastage by allowing the operator to prepare future jobs while running current jobs. It will handle queues of programs, make sure that all necessary devices are available, and attend to various housekeeping duties, such as blocking and unblocking records on magnetic storage. This aspect of file control, which is part of job control, was discussed in chapter 5. Another task it can perform is insuring that the correct files are available for a particular job, and that the operator has not loaded an incorrect one. In an organization with scheduled work, job-control software can be instructed each morning as to what jobs are to be done, and the software can then control the work.

Common tasks 13.2

It was realized soon after computers became of commercial significance that certain tasks had to be performed by every user. Because it was wasteful for programmers in each installation to write programs for these tasks, the manufacturers supplied software to perform them. Common-task software can be subdivided into organizational, system, and programming software.

Organizational common tasks are those the user can select for his particular business. In this category are the standard routines dealing with FICA. The distinction between these routines and those that will be considered later as application packages is that application packages are complete in themselves. A routing for calculating FICA, on the other hand, is designed to be incorporated as part of a larger payroll program. The software simply saves the user programmer the labor of flowcharting and coding a standard calculation.

Common task programs are sometimes called utilities. The largest selection in this group perform sorting. As discussed in chapter 5, a number of techniques are available for sorting data into a prescribed sequence, and all of them are available as standard software on various machines. Other utility programs are report generators that help the programmer produce intelligible reports according to his layout.

In the programming area, common-task software usually consists of subroutines. Sets of coded instructions that the programmer can insert into his own program are supplied. In this respect, they are closely linked with the organizational common-task software. However, organizational software is normally associated with a particular application such as payroll, whereas programming software covers tasks that are common to all applications. Examples are the subroutines that calculate square roots, perform iterations, or add days to the date. Many statistical subroutines such as calculating standard deviations are in this category as well.

13.3   Programming aids

Another major group of software is designed to assist in the actual task of programming a system. This software may be subdivided into test systems and languages. The first group includes all specialist software enabling the programmer to discover logical errors in his coding. It has been said that the only thing all programs have in common is the certainty that they will not work the first time. The programmer spends a significant proportion of his time testing small sections of his program to correct inconsistencies. When this job is finished, he must link the tested segments to make an operational program. Finally, he must prepare test data that will insure that all paths through the program are proven. At each of these stages, the test systems software is helpful and usually reports actions that have been performed and errors that have been found.

The other subgroup of programming aids—languages—consists of a number of levels. The subject of software languages is very diverse, and many different types have evolved. However, all languages can be classified into three main categories: machine, low-level, and high-level languages. Their purpose is common: to enable the user to communicate with the machine.

The computer hardware is activated by a series of impulses in binary form. To program a task as a set of binary coded instructions is extremely difficult and error-prone. When it is done, the resulting program is said to be written in machine language, because it bears a close resemblance to the instruction patterns executed by hardware. Today programs are rarely written in machine language. Such programming is done only when a special task cannot be accomplished by other means.

Following closely behind machine languages came a large number of low-level languages. Every new machine had its own language. With a low-level language, each function that the machine is capable of performing is given a mnemonic name that the programmer uses when he requires that particular function. For example, he might write SUMSQ when he wants to "sum the squares of," or XECA to "execute the instruction in location A." The computer accepts these mnemonics and translates them into appropriate machine codes. These low-level languages assist the programmer, but he is still required to develop the logic at the machine level and consider the steps in operational sequence. The use of the language helps primarily in the coding operation.

This restriction led to the development of various high-level languages which were problem-oriented rather than machine-oriented. Their underlying principle is that the user understands the problem he wishes to solve via the computer and can write it in a form that closely resembles English. There are certain rules that must be learned before the high-level language can be used, but in most cases they can be learned very quickly.

The high-level languages in common use are:

1. FORTRAN, which was designed primarily for expressing mathematical formulas easily. The language is available in different versions, of which FORTRAN II and FORTRAN IV are the most widely used.
2. ALGOL, which is another mathematical language. It is widely adopted in manuals to document programs and in subroutines to perform advanced mathematical techniques.
3. COBOL, which is basically designed for commercial applications. It has good facilities for file processing and input/output operations. As with FORTRAN, more than one version exists; each is a subset of the complete language.
4. PL/I, which is a language that combines both mathematical and commercial facilities.

Because of basic differences in the design of computers, it is not always possible to incorporate all the facilities of a language into the program that translates the high-level language into machine code. For this reason, it is not safe to assume that a program written in a high-level language for one computer can be run on another machine even if that particular language software is offered on the second machine.

13.4   Application packages

The application package is a type of common task program that is sufficiently important to be considered in its own right. It is designed to avoid duplication of programming effort and tends to be self-contained. It has standard input and output documentation that makes it easy for the user to apply in his business. Sometimes it contains an option under which the user can link his program to it. For example, an organization may wish to use a package as an integral part of a larger system. In this case, the normal printed results provided by the package may not be required, and the user can add his output module to replace the print segment in order to collect the necessary data on magnetic tape.

Some application areas that have been covered with packages are:

1. General purpose simulation systems (GPSS)
2. Automatic type justification
3. Business information systems and information retrieval
4. Chemical engineering
5. Civil engineering
6. Electrical engineering
7. Inventory management
8. Numerical control
9. PERT and critical path scheduling
10. Profit evaluation
11. Traffic engineering
12. Payroll
13. Production control

In keeping with other computer fields, mnemonics abound in this area. Names such as PERT, ECS, ADS, ADAPT, and many others appear frequently in the literature.

A considerable number of application packages are based on applied statistical methods and can have a marked effect on management methods. Decision-making is often assisted by correctly applying appropriate statistics. The small computer user can apply sophisticated techniques without employing highly specialized analysis and programming staff. Other effects of the use of standard packages are the discipline that is enforced upon input procedures, the greater confidence in computer output, the closer liaison with the manufacturer, and the speed with which special applications can sometimes be performed in a particular area of the business. Without these packages, some areas of investigation would be ignored indefinitely because of the programming effort required.

Application packages have limitations, however. Sometimes the user's system may have to be modified to tailor it to the package's input requirements. This may cause problems, since with any computer system, the output will only be as good as the input. The most sophisticated mathematical techniques will produce incorrect answers if fed ambiguous and invalid data. The analyst must also beware of the programming implications of any proposed package. Because of their generalized nature, some packages use more computer time and storage than would be needed by a customer-built program. Occasionally, this factor may make it advisable to write a program instead of buying one.

Exercises

1. What two main functions does software perform?

2. Why is an operating system so important in a multi-programming environment?

3. What types of activities does an operating system perform? Give an example of each.

4. Contrast organizational, system, and programming software, and give an example of each.

5. In what way does a test system aid the programmer?

6. If high-level programming languages are so much easier for the programmer to use, why do low-level languages even exist?

7. Contrast the four most commonly used high-level languages.

8. Give three examples of application programs. What advantages and disadvantages does this type of program offer over a user-written program?

appendix a

information retrieval

The analyst must recognize that the term *information retrieval* may have two different but related meanings. The first is applied almost exclusively to document retrieval, and refers to the selection of published information. The second use of the term describes the selection of relevant information from a data bank of current company operating information. Both operations can be performed by manual methods, and both can be widened in scope by using a computer.

The function of a computer-based information retrieval system is to identify records relevant to a search request specified in computer terms. Such a definition applies equally well to both types of problem outlined above. The fundamental difference lies in the basic computer files involved. For document retrieval, the files have a static information content. Once recorded and correctly indexed, the book, periodical, or document is available in the same form until it is physically withdrawn from the library. With commercial retrieval systems, the opposite is true: the files are dynamic and their contents change as each commercial transaction is recorded. This is one reason why applications of true information retrieval are very limited in the commercial field. Extensive work is being done on document retrieval methods by libraries and scientific professions.

Most effort in the document retrieval field has been devoted to improving indexing techniques. One of the earliest uses of computers for this task was to produce Key Word In Context (or KWIC) indexes. These select all indexable words from the title of a document and sort them so that each word becomes an index entry in its own right. The remaining words of the title appear in their normal context, usually to the left and right of the indexing word. After sorting is completed, the index is printed in alphabetical sequence of the index words.

A development of the KWIC method was Selective Listing in Combination (SLIC). Under this system, the indexable words within a title are combined. Each combination that is logically contained in a more specific combination is omitted. The SLIC system also restricts the number of indexable words in any title to a maximum of five, generating a total of sixteen entries when all five are used.

A third approach was the introduction of chain indexing. This method is similar to SLIC but the indexable words are considered to be chained to each other. It is necessary to produce only as many entries in alphabetical order as there are indexable words in the title. Each entry omits the leftmost word from the previous entry in the chain and starts with the full title. Figure A-A compares the SLIC and chain methods.

These methods can generate a computer-printed catalog of all titles currently published within any chosen subject field. The user can then consult the index to see if any material is available dealing with the subject of his interest. In the scientific field, particularly, so much material is published each year that even the task of scanning a comprehensive index can be burdensome.

To overcome this problem, computer specialists and librarians introduced a system called Selective Dissemination of Information (SDI). Under this scheme, each user presents

ORIGINAL ARTICLE

	Document Ref: IBJ 484-116
Title: Programs for Generating Random Numbers	

SEQUENCE OF DESCRIPTORS CHOSEN FOR INDEX

RANDOM	NUMBER	GENERATING	PROGRAMS

SLIC INDEX ENTRIES

RANDOM NUMBER GENERATING PROGRAMS	IBJ 484-116
RANDOM NUMBER PROGRAMS	IBJ 484-116
RANDOM GENERATING PROGRAM	IBJ 484-116
RANDOM PROGRAMS	IBJ 484-116
NUMBER GENERATING PROGRAMS	IBJ 484-116
NUMBER PROGRAMS	IBJ 484-116
GENERATING PROGRAM	IBJ 484-116
PROGRAMS	IBJ 484-116

CHAIN INDEX ENTRIES

RANDOM NUMBER GENERATING PROGRAMS	IBJ 484-116
NUMBER GENERATING PROGRAMS	IBJ 484-116
GENERATING PROGRAMS	IBJ 484-116
PROGRAMS	IBJ 484-116

Fig. A-A. Comparison of SLIC and chain index methods

an interest profile to the indexing center, stating subject headings in which he is interested. As new accessions are made to the main index, each entry is compared by the computer against the stored profiles. When a match between an indexable word and a subject heading in a profile is obtained, a copy of that index entry—and sometimes an abstract of the original document—is sent to all interested users. More advanced systems can provide microfilmed copies of the appropriate document instead of an abstract.

An obvious limitation of the systems described above is the use of the document's title to form an index entry. In scientific works, this problem is reduced because titles tend to be long and fairly descriptive of content matter. In non-scientific fields, many titles are ambiguous or simply devices to catch the reader's eye. For example, a recent article describing hybrid computers was entitled "The Vital Link." Whatever indexing system was used, it is unlikely that the article would be successfully retrieved in response to a request for information on hybrid computers.

A solution is to select descriptors for indexing from the title, if possible, or from an abstract of the contents when the title is not informative. A disadvantage of this scheme is that all incoming documents have to be abstracted by competent specialists before they can be indexed. Consequently, much of the speed and automatic nature of the computer system is lost.

This raises another problem that is crucial within any document retrieval system: the user has no prior knowledge of the records he wants. Normally he will not be the person who allocated the descriptors, and his interpretation of words is likely to be different. How can he select the most suitable words to express his request? One solution is to require the user to frame his question from the words contained in a thesaurus or a lattice. A thesaurus links ideas with specific descriptors. The user can then consult the list and choose the descriptor that most closely resembles his idea of the subject he wishes to retrieve. A lattice is a set of words in which all useful interrelations between all words are defined. This is a totally different kind of language and is harder to maintain, although more sophisticated. Much work must go into the planning of the system. Instead of arranging all information in a conventional hierarchical structure, a lattice contains all links between individual items.

Another solution is to let the user find as many synonyms as possible for the descriptor of current interest so that the index can be searched for each of these if a main word does not select sufficient relevant material. By far the most ambitious scheme is to store a complete abstract of the material, thus increasing the number of descriptors available and giving the user an almost infinite choice. This latter method, which is the subject of experiments in some disciplines, is usually called free language searching.

A wide variety of other ideas, such as the use of standard orders of priority to determine the logical sequence of words, and the use of weights given to certain descriptors for the same purpose, are in a fairly advanced state of development. They fall far short of the ideal of a totally free language system. However, free language systems and systems that automatically split sentences into their syntactical structure and words into their philological structure are now being developed with considerable success.

For all the indexes considered so far, an index entry is necessary for every record contained within the system. Such indexes tend to be bulky and inefficient for machine searching because they involve sequential file-reading techniques. An alternative approach to indexing the data on file is known as a coordinate index. The manual versions of this method are often called feature card systems.

Coordinate indexes contain an index for each descriptor, or for each descriptor value, giving all the records containing the descriptor. Such a file is called an *inverted* file. The retrieval process is considerably faster than in a sequential file, since only those documents known to contain the desired descriptor are accessed.

A serial file lists each document by number and includes all descriptors relating to that document. Thus, to locate all documents with (a) given descriptor(s), it would be necessary to search the entire index serially, examining each document for the presence of the desired descriptor(s). The advantage is that any document can be located quickly if its number is known. The disadvantage is that the retrieval process is otherwise slow, and it is therefore not generally used for large files.

An inverted file lists each descriptor and includes all document numbers containing that descriptor. The advantage is that the retrieval process is considerably faster, since only those documents known to contain the desired descriptor are accessed. The disadvantage is that it is difficult to limit the search to those documents containing a certain combination of descriptors; a search must be made for each descriptor, and the several lists of documents thus retrieved must be cross-referenced to distill out those documents with the desired combination of descriptors. Such a system will reveal which records contain all specified descriptors. If we wish to know which records contain specific combinations of descriptors, the retrieved documents may be cross-referenced and indexed in the same way. There are techniques for packing and processing such large indexing systems very efficiently. Updating and using such indexes requires extensive software.

Figure A-B compares aspects of sequential and inverted

file techniques. The final choice depends on system requirements.

A variation of this form of inversion is the chaining method
now developing in several advanced retrieval systems in
commercial use. Each descriptor belonging to a given record
can possess a subset of descriptors, and so on, to a depth of
a dozen or more. Reference and information retrieval is still
achieved by indexing each category of descriptor, but each
descriptor record has references to the adjacent descriptor
record in the same chain, and to the master descriptor record
in the chain that belongs to the chain above. The system
will make commercial aspects of information retrieval stronger
and will allow more powerful file-maintenance techniques.

In the commercial field, information retrieval in its narrow-
est sense simply means conventional file interrogation. For
example, the problem might be to find the record for the

TABLE A-B

SEQUENTIAL AND INVERTED FILE COMPARISON

COMPARISON	SEQUENTIAL FILES	INVERTED FILES
Updating	Only one record has to be accessed and changed; fast-est	Several records have to be accessed and changed. Slow-er and more complex
Record format	All data in a record is in one place; each record remains essentially constant after entry.	Single records may be sepa-rated, and extensive cross referencing, or indexing, is required. Each record is con-tinually changed.
Searching	(Much) slower	Fast, especially with direct access files
Hardware	Suitable for tapes	Suitable for disks, drums, or magnetic card files
Extra space occupied	None	Can be extensive and must be random access
Logic	Most logical operations are easily fulfilled.	"OUT OF" and "BETWEEN" are difficult, and take up much time and working space: the amount of stor-age required is doubled for every level above two.
Suitable file size	Adequate up to 20–30,000 records	Starts producing benefits at about 20,000 records, assum-ing an adequate request rate.

customer who has account number 123 and print out his current ledger balance. The kinds of indexing and semantic problems that document retrieval has do not arise here. However, more sophistication is added when records that satisfy more than one criterion are requested. Semantic problems are still minimal, but search strategies can become complex.

Three examples of requests that might be made in a commercial system are:

1. How many properties in the file have more than three bedrooms? In this census type retrieval, it is assumed that details of each property are available on the file. One field of each record contains the number of bedrooms. To extract the required information, the file is searched and a count is made of the records for which this particular field has a value of four or more.

Two points mentioned earlier can be illustrated by this inquiry. One is the semantic problem: does the descriptor "properties" really mean what it says, or was the inquirer referring to houses, and excluding hotels, residential schools, and so forth, which might be on the file? The second point is that if a coordinate index were available for the file, and descriptors had been allocated for "1 bedroom," "2 bedrooms," and so on, the inquiry could have been answered directly from the index without searching the file.

2. In the northeast region, which retail outlets, excluding those that receive a special discount, have sold more than 4,000 units of product A?

This extension of question (1) requires combinations of fields on each record to be examined to match with the inquiry.

3. What will be the overall effect on our manufacturing costs of an increase of $5.00 per sheet in the price of metal of specification Z?

This type of inquiry requires more than a simple record match, although this must be the first step. The raw material file must be searched to determine which products use sheet metal of this specification. During the same search, the number of square feet of used metal must be extracted and totaled. The final total divided by the number of square feet in a sheet and multiplied by five is the total price increase in dollars. Subject to any special purchasing information that might be needed (for bulk discounts or minimum quantities), a final step might be to express this result as a percentage of the total manufacturing costs. The reply might then be made that costs would rise by eight percent.

Systems considerations in designing a file for these types of inquiry will include the usual problems of field structure, data management, and so on. They will also include access rates, file-maintenance activity, the distribution of the number of descriptors over the number of records, the characteristics and the homogeneity of the descriptor collection, the indexing mechanism, and the depth of logic required.

An information retrieval system consists of more than just a set of files and indexes. It contains a complete coding and conceptual structure that implies a total formalization of the subject under scrutiny. The outward signs of this are a dictionary of terms used in the system, a code representing the terms, and instructions for the use of them. But this is deceptively simple, because before the system is used, every item in the fields covered must be totally defined and correctly coded. This may involve years of work, whether the object is to get computers to "understand" words, to identify any chemical substance by either its structure or its name, or to identify any piece of commercial data that is relevant to a marketing man's inquiry.

The development of a genuine information retrieval language to suit a given application, excluding the data-processing aspects, can take many months. This emphasis is reflected in the literature on information retrieval, which has currently ten percent devoted to mechanization techniques and ninety percent to semantic development. The effort is devoted to insuring that the words used and their coding are sufficiently complete to be useful to all workers in the field carried by the system.

In conclusion, two concepts should be mentioned. *Relevance* is the percentage of the retrieved items that are relevant. *Recall* is the percentage of relevant items that were retrieved. These concepts conflict, and a deliberate compromise between them must be reached. On the one hand, if all items, including those even remotely related to the inquiry as phrased, are retrieved, the inquirer will waste much time reviewing and rejecting inapplicable items. On the other hand, if only those items closely related to the inquiry are retrieved, the inquirer may miss many items of interest, either because of the phrasing of his inquiry or because of poor cataloging procedures. The compromise between recall and relevance is often the success criterion of an information retrieval system.

appendix b

systems and information theory

In his practical day-to-day activities the analyst will have little opportunity to examine the theory underlying systems design. He will find himself far too busy with the details of his system to be able to spare time for such academic pursuits. However, since practical work is based on theoretical foundations, this section will explore some of this theoretical ground to give the analyst a basis for further study.

Attempting to define the word "system" in the sense in which it is used in such prases as "systems analysis" or "systems engineering" is difficult, perhaps even impossible. Many feel that such an attempt is doomed to failure. The word is so general that any attempt to define it requires the use of other words and these, being even more general and inclusive, are less well understood than the word "system" itself. It is impossible to define all the words in a language without becoming circular. Perhaps the word "system" could quite properly be classed as one which should be left undefined; its meaning should be acquired by observing it in use. However, it is possible to make statements that can be indicative if not actually informative about things considered in systems analysis.

The reason for making this attempt at definition is that the word "system" is used in everyday language with at least two meanings that disagree with the use implied in the phrase "systems analysis." In common language, the word is sometimes used to refer simply to a collection or aggregation of similar or interrelated things such as "a solar system," or "a system of numbers." It is also used to mean a collection of rules for procedure such as "a system for winning at roulette." We are concerned with the use of the word applied to a situation in which both elements are present. A system, therefore, will consist of a collection of interrelated things together with a set of rules for procedural behavior. The collection of things is called the system composition and the set of rules is called the systems operation. There is a third requirement for systems to be subject to analysis: either the composition or the operation of the system or both must be under human control. The solar system is not subject to systems analysis, but an artificial earth satellite is.

A system is a collection of entities or things (animate or inanimate) that receives certain inputs and is constrained to act on them to produce certain outputs, with the objective of maximizing some function of the inputs and outputs. The essential characteristic of a system is its connectedness. That is, anything that consists of parts connected together can be called a system. For instance, a game of checkers is a system, whereas a single checker is not. A car, a pair of scissors, an economy, and a language are systems. They are aggregates of bits and pieces, but they begin to be understood only when the connections between the bits and pieces and the dynamic interactions of the whole organism are studied.

It will be noticed that the definition uses some very general words indeed—entities, inputs, outputs, function. Nonetheless, some of the important distinguishing aspects of things treated in systems analysis are suggested by this collection of words.

First, the words "act on" are very important. A system is something dynamic. A completely static object is not a system. Thus for most people a stone is not a system, although to a geologist who sees a stone as something that used to be mud and is in the process of becoming sand, a stone might be, at least, dynamic. A building, as such, is not a system. A hotel, together with its staff and operating rules, which receives inputs (food, fuel, guests, complaints, water, bills) with the objective of maximizing profits is a large and complex system. It contains many subsystems: a heating system, a plumbing system, an accounting system. If we wish to consider the interactions that affect one single entity, then we shall have to define that entity as part of a system. The system we choose to define is a system because it contains interrelated parts and is in some sense a complete whole in itself. But the entity we are considering will certainly be part of a number of such systems, each of which is a subsystem of a series of larger systems.

Next, the words "with the objective of" suggest an important fact. For our purpose, a system exists only when somebody has something in mind: there must be intent. A thunderstorm receives inputs and produces outputs; it is undoubtedly dynamic. The dynamics of a thunderstorm can be and have been subjected to mathematical analysis. And in common language a thunderstorm is often called a system. However, the lack of human intent or control of the dynamics involved makes most of the techniques of systems analysis inapplicable to thunderstorms. Accordingly, the words "systems analysis" are usually restricted to situations where the interaction between human intention and actions and the performance of the system are being considered. Finally, consider the way in which the "objective" of a system was stated, that is, "maximizing some functions of the input and outputs." It may be felt that this is rather special and restrictive, but this is not the case. Any imaginable result that depends on values of the inputs and outputs can be expressed in this way.

The word "function" is very general indeed and simply

means something with a value that depends on the inputs and outputs. The function to be maximized may be thought of as some measure of worth or value of the system. The process of systems optimization actually consists of two parts:

1. The formulation of a (value) function which it is desired to maximize, and
2. The variation of the systems composition or operation in such a way to accomplish the maximization.

The first of these problems is often the more difficult. Let us assume that we have succeeded in isolating and describing the system with which we wish to deal. Let us now represent the "bits and pieces" that make up this system by a series of dots on a paper. The connectiveness of the system can be introduced by drawing lines between the dots; some dots may be connected to all other dots, but in some cases a dot may be connected to only one of its fellows. In this way, we can see a system as a kind of network. The feature of this network in which we are interested is the pattern created by the lines. It will probably change from moment to moment as the system interacts within itself to operate in its own way. The nature and the extent of the control the system displays is revealed in the behavior of the pattern of this network.

The task of studying a system is therefore a rather special one. A system consists of n elements. Without considering interconnectedness, this would have meant n investigations to find out what this collection of things was like. Once we say the set of things is a system, there are not only the n elements themselves to examine, but also $n(n-1)$ relations between the elements. If a system has only seven elements, it has forty-two relations within itself. If we define the state of this system as the pattern produced in the network when each of these relations is either in being or not in being, there will be 2^{42} different states of the system. This is more than four millions of millions! This is the basic and formal reason why the rigorous and exhaustive study of systems is so difficult. A system in a dynamic state may pass quite rapidly from one state to another for an indefinite period. Accounting for this behavior will obviously require a vast investigation.

In discussing "objectives," the term "optimization" was used. The process of optimization always takes place with a number of restrictions depending on what variables of system composition or operation are varied. Hence every optimization is really a suboptimization. For example, consider a thermostatically controlled heating system in a building. A heating engineer called in to adjust it will only look at minor variables (voltages, contact spacing, and so forth). His optimization will consist of minimizing the difference between the temperature called for (TC) and the temperature actually provided (TP). A heating engineer designing a system for a new building will certainly not attempt to minimize the difference (TC-TP), as his system value. An engineer developing a heating system for general public sale must also consider customer appeal and will have to include more complex factors such as aesthetics or noise in his value function. Finally, from a broader point of view, the "optimum" solution of the heating problem might be to breed a new race of man impervious to temperature variations and so eliminate all the mechanisms. This is what is meant by saying that in the real world only suboptimizations are performed.

It has been stated that systems analysis applies to situations in which one is considering the interaction between human intentions and the performance of the system. It is convenient to classify systems according to the way in which this interaction enters the problem.

The minimum human involvement occurs in a system whose components are purely mechanical (chemical, electrical, hydraulic, pneumatic) devices and whose operation is determined by the laws of physics and chemistry. A guided missile is such a system. The human element is confined to the design of the system composition. This activity is pure systems engineering, and such a system can be called *mechanistic*.

Many systems are designed to operate with a set of important inputs so numerous that it has not been possible to create a set of operating rules to cover all possible variations of all input parameters. For example, a manufacturing organization must be prepared to cope with variations in raw material supply and cost, in consumer demand, in the tax structure, and in a host of other regulations, in population, wages, unions, transportation, and so forth. Such a system uses humans in an essentially nonmechanical way to make decisions that modify system operation. Since the purpose of the modification is to adapt the system operation to a new set of conditions, such a system is often called *adaptive*. The techniques for analyzing an adaptive system are quite different from that applying to a mechanistic system.

Finally, one specific type of variation of input poses problems of such a special nature that it is necessary to distinguish this situation and develop special techniques for its analysis. This situation arises when the system must operate with other systems whose objective is to negate that system. Commercial competition and war are the two most common areas in which these situations arise, and such systems are called *competitive*.

It will be useful to base an arbitrary classification of systems on two distinct criteria. One obvious criterion is system complexity. According to this criterion, systems may be discussed in three ways. The least complex may be called *simple but dynamic*. A system that has become highly elaborate and is interconnected will be called *complex but describable*. Third, we may discuss systems that have become so complicated that, while they may be designated as complex, they cannot be described in a precise and detailed fashion. Such systems will be called *exceedingly complex*.

The second criterion on which this classification will be built concerns the difference between *deterministic* and *probabilistic* systems. A deterministic system is one in which the parts interact in a perfectly predictable way. There is never any room for doubt: given a last state of the system and the program of information defining its dynamic network, it is always possible to predict its next state without any risk of error. A probabilistic system, on the other hand, is one about which no precisely detailed prediction can be given. The system may be studied intently and it may become easier to say what it is likely to do in any given circumstances. But the system is not predetermined, and a prediction about it can never escape from the logical limitations of its probable behavior. A sewing machine is a deterministic system: you turn the wheel and the needle goes up and down. A dog acts in most cases as a probabilistic system: you offer it a bone, and it is likely to come towards you—very likely. On the other hand, it may suddenly run away.

These two criteria, one three-fold (simple, complex, exceedingly complex) and the other two-fold (deterministic, probabilistic), produces a classification system of six categories.

A simple deterministic system has few components and interrelations and reveals completely predictable dynamic behavior. The door latch on a car door is an example. A force applied to the handle disengages the latch from its catch: that is all there is to that system's operation.

Equally, a game of billiards, if defined appropriately, belongs to this class. Billiards is a system of simple dynamic geometry. It is entirely deterministic, because a given shot always has a fully predictable effect. This sytem becomes probabilistic only if it is adapted to describe an actual game. Then the imperfections in the table, balls, and cues, not to mention the skill of the players, will introduce so many variables that the system will become probabilistic.

The same careful distinction is needed to evaluate a third kind of a simple deterministic system, the layout of a machine shop. A problem of this kind may be studied in terms of the movement of material along a predetermined route. Distances material must travel may be minimized within such a system. When what actually happens when material begins to flow is studied, the system at once becomes probabilistic. This example is similar to that of the billiards game: while it remains an abstraction the system is deterministic; it loses this character once it is restricted by real life.

Similar considerations apply to complex deterministic systems. Instead of a window catch, we might consider an electronic computer. No one who has looked inside one of these machines would call it simple; it is an extremely complicated mechanism. On the other hand, it is entirely deterministic. A computer will do only what it is precisely told to do. Insofar as its behavior is not fully predictable, it has been inadequately directed. Thus an electronic computer as we know it today is an excellent example of a complex deterministic system.

For a second example, the billiard table may be enlarged to cosmic proportions. Consider the behavior of the visible universe. The search for "laws" in this system has been sufficiently successful; the observable movements in the heavens are predictable. If in any degree they are not, the scientist merely declares a small gap in his knowledge and seeks to fill it. The slight variations in the predicted movements of orbits, for example, have not been ascribed to their probabilistic nature; on the contrary, they have been used as a basis for the formulation of further hypotheses. Such a procedure as an actual method of science is only reasonable insofar as the system may be assumed to be deterministic.

Turning once more to industry for a third illustration of the complex but deterministic system, we may consider the automated factory. Again we have a complex system. And again any deviation from a fully predicted course on the part of, say, a battery of transfer machines would be regarded as a breakdown. The regulation system is automatic, deterministic.

Now consider these two levels of complexity when the systems involved are not deterministic. First, there can be a simple probabilistic system. For example, consider the tossing of a penny. Here is a perfectly simple system, but a notoriously unpredictable one. It may be described in terms of binary decision process, with a built-in even probability between the two possible outcomes. For another illustration in the simple category, consider the behavior of the jellyfish. Here is a remarkably simple organism, but one whose movement, even in still water, is not predictable except in terms of mathematical probability. (Incidentally, considered biochemically, the jellyfish is exceedingly complex. Note how vital it is to prescribe what aspect of a system is being considered.)

For an example of the complex probabilistic system, think of stock. This stock may be on the shelves of a retail shop, raw material inventory, or the inventory held against production breakdown. This is a physical system: things arrive and are put into stock, a call for something arises, and things are withdrawn from stock. How far can we go in describing this as a system? A very long way, because both the arrival process and the departure, or service, process are describable in terms of mathematical statistics. That is, they are probabilistic in nature, but they can nonetheless be described. Even when the stock-holding situation is very complex, with a great many sources of input and of demand for output, the system can be treated within this category.

A second example from industry is the concept of profitability. The system for making a profit in a manufacturing situation is essentially a complex probabilistic problem. A change is made: perhaps it is a simple move to increase profit by changing a single product. But almost certainly this change belongs also to a bigger system of the kind we are now considering. Making the change scatters influences in all directions; what these influences encounter and how the whole system is modified until it settles into a new balance cannot be exactly predicted. The final effect on profitability is, the management hopes, an overall gain. But the scientist, if he can estimate it at all, must do so through the theory of probability.

So far, this discussion has covered four of the six categories mentioned above. No mention has been made of the exceedingly complex systems, which have been defined as so complicated that they are virtually indescribable. For this reason the category of exceedingly complex deterministic systems is declared empty. Any fully deterministic system, such as the astronomical system considered earlier, can eventually be described in detail. However complex the system may become, it will in principle be possible to specify it completely. Thus it is claimed that there are no members of the exceedingly complex class in the deterministic category.

In the exceedingly complex probabilistic category, however, the story is very different. The country's economy, for example, is so complex and so probabilistic that it seems unreasonable to imagine that it will ever be fully described. Another, living example—the human brain—can also be described in this way. Moreover, the brain is notoriously inaccessible to examination. Once it is dead, it is no longer a dynamic interacting system, and in any case it disintegrates rapidly. Surgical examination is not very helpful and is extremely risky. Inferential investigations about its mode of working, from such studies as psychiatry, are progressing slowly.

Probably the best example of an industrial system of this kind is a business organization. It is certainly not alive, but it has to behave very much like a living organism. It must develop techniques for survival in a changing environment: it must adapt itself to its economic, commercial, social, and political surroundings and it must learn from experience. Figure B-A summarizes the examples discussed above.

TABLE B-A
BROAD SYSTEMS CLASSIFICATION

SYSTEMS	SIMPLE	COMPLEX	EXCEEDINGLY COMPLEX
Deterministic	Door latch	Electronic computer	(None)
Deterministic	Billiards	Planetary system	(None)
Deterministic	Machine shop layout	Automation	(None)
Probabilistic	Penny tossing	Stockholding	The economy
Probabilistic	Jellyfish movements	Conditioned reflexes	The brain
Probabilistic	Statistical quality control	Industrial profitability	The company

One of the specialized subjects relevant to the understanding of complex systems is information theory. This new area of study owes its origin to the work of communications engineers and is growing rapidly because of the growth of information technology. It is associated with the interest in control systems and the broad, if somewhat diffuse, area of cybernetics. A theory of communication is concerned with carrying information from one location to another with little reference to its meaning. It is thus concerned with communication channels and the transmission of messages from a source to a destination. In the context of the work of the communications engineer, the theory can help in designing the optimum equipment. In the wider context of business systems, the theory cannot give precise rules and formulas for solving systems design problems. It can, however, provide a basis for understanding these problems and help to clarify the nature of information. The remainder of this section introduces some of the basic concepts of information theory.

Figure B-B is a simple diagram of general communication channels. The source generates a message, which, via the coding operation, becomes a signal. The signal received, which may not be exactly the signal sent, is decoded to give a message at the destination. The concept behind this arrangement is fundamental to statistical information theory, in which information can be considered as a quantity. This implies the existence of a predetermined code; information is then a quantity of selections from among the possibilities presented by the coding system. It is in the form of a number of bits selected from a well-defined code or "alphabet" for transmission along the channel. This concept has the added convenience that, technically, discrete signals are preferred over continuous signals. Complex codes, such as Morse code, can be built with a few readily distinguishable discrete signals, and the receiving instrument need only decide which permitted signal occurred. This is very relevant when interference, or noise, is acting on the channel. The particular sequence or order of the coded signals determines the meaning of the message, which involves a relationship between the message source and the recipient, wider than the mere connection of the communication channel.

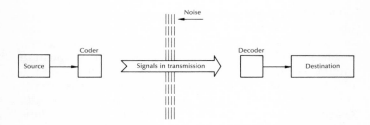

Fig. B-B. Communication channels

To arrive at some way of measuring the amount of information in a message, consider a simple coding system. It is convenient to postulate a noiseless channel—that is, one in which there is perfect reception of the signal as sent—and a system that can transmit only two signals—0 and 1. If for simplicity we only have four messages W, X, Y, and Z, then we could code these as follows:

Message W as O, and use one binary digit or "bit."
Message X as 1, and use one bit.
Message Y as 00, and use two bits.
Message Z as 01, and use two bits.

If these messages occur equally among the messages received, then the probability of any one occurring is ¼. The average bits per message is then:

$$(1 \times \tfrac{1}{4}) + (1 \times \tfrac{1}{4}) + (2 \times \tfrac{1}{4}) + (2 \times \tfrac{1}{4}) = 1\tfrac{1}{2} \text{ bits}$$

We cannot reduce this average unless the messages occur with unequal frequency. If this is the case, we can improve the efficiency of the system, in terms of bits needed per message, by assigning the shortest codes to the most frequent messages. For example, if—

W occurs once in every 8 messages, code it 00, using 2 bits

X occurs once in every 8 messages, code it 01, using 2 bits

X occurs twice in every 8 messages, code it 1, using 1 bit

Y occurs four times in every 8 messages, code it 0, using 1 bit

The average bits per message is now improved to:

$$(2 \times \tfrac{1}{8}) + (2 \times \tfrac{1}{8}) + (1 \times \tfrac{1}{4}) + (1 \times \tfrac{1}{2}) = 1\tfrac{1}{4} \text{ bits}$$

The basis for measuring the amount of information may limit the improvement we can make in coding. With a larger number of messages than 6, we shall have to go to three bits for some messages, and so on, so that we may end up with a wide range of bits per message. Consider the problem of computing the amount of information in a written sentence. This problem of great interest and importance is not yet solved because of the lack of complete statistical data about the language, even though the probability of occurrence of letters in English can be established.

It is possible to look on the amount of information in a message as proportional to its surprise or news value, so that the least frequent message has the greatest impact. A measure is then given by the reciprocal of the probability. In order to obtain the desirable property for the measure so that we

can add the individual information values of a number of
discrete signals, we take the logarithm and postulate

$$I(m) = \log \frac{1}{P(m)}$$

where $I(m)$ is the amount of information received in the
message m, and $P(m)$ is the probability of this message occurring. The base of the logarithm is the unit of information,
thus —

1. Base 10 for decimal digits
2. Base 26 for alphabetic symbols
3. Base e for natural logs
4. Base 2 for binary digits or bits

In computer techniques, it is usual to employ the binary
scale and construct the logs to base 2 so that units of measure will be bits. If we take S as the number of symbols in
the groups, so that $S = 26$ for letters, or 10 for decimal numbers, then the amount of information is $\log_2 S$ per character.
A message which consisted of 4 decimal digits and 2 letters
could then require —

$$4(\log_2 10) + 2(\log_2 26) = 22 \times 7 \text{ bits}$$

This is a useful concept of measurement within the bounds
of the communication system we are considering. It shows
that the information capacity of a character depends on the
extent of the group from which it is drawn, and the group
must be fully specified beforehand.

The *efficiency* of a code is defined in standard terminology
as

$$\frac{x(\log_2 n)}{H(S)}$$

where $H(S)$ is the actual average amount of information, x
is the length of the code, and $\log_2 n$ is the maximum information capacity of one character of a set of n characters.
Redundancy is then defined as —

$$(1 - \text{Efficiency})$$

and should be regarded as an addition to the essential information. One such example is the use of parity bits. An-

other is the vowel character in written language. Except at the beginning of a word, vowels can often be omitted and the word still understood.

Noise is an interference that destroys some part of the signal and initially at least may be considered random. The noise effect introduces uncertainty about the message. The average uncertainty per character is termed *equivocation* and gives a measure of the amount of information lost in the channel.

If complex coding is employed, using the latent possibilities of the redundant element of a code, it is possible to communicate through a noisy channel with less risk of error. Shannon's second theorem states that "for any information rate less than the channel capacity, a code can be found to make the loss of information as small as desired."

Entropy is a term that originated in the study of heat. Broadly, the flow of heat from a higher-temperature body to a lower is considered an increase of entropy. It is a natural process of nature, so any isolated system or "universe" tends to equalize temperature and maximize entropy. Such a universe is increasingly random and uninteresting because it lacks organization. Local centers of organization, which are effectively pushing against this tendency towards equalizing, represent reduction in entropy and increase in variety. Such an effort is provided by information. Entropy is a property of the arrangement of the parts of a system, and decreases as the arrangement becomes more distinguishable, that is, as the information we have about the system increases. Information is thus the negative of entropy.

This introduction to information theory may seem too theoretical to be of immediate relevance to the systems analyst. However, it has covered a number of items increasingly in use, and a number of concepts fundamental to the coding and transmission of information. Its broadly educational approach will, it is hoped, stimulate interest and encourage further reading while its specific training implications will be apparent.

appendix c

selected readings

Arkin, H. "Computers and the Audit Test," *The Journal of Accountancy,* October 1965, p. 44.

Blumenthal, Sherman C. *Management Information Systems: A Framework for Planning and Development.* Englewood Cliffs, New Jersey: Prentice-Hall, Inc., 1969.

Boguslaw, Robert. *The New Utopians: A Study of System Design and Social Change.* Englewood Cliffs, New Jersey: Prentice-Hall, Inc., 1965.

Boutell, Wayne S. *Computer-Oriented Business Systems.* Englewood Cliffs, New Jersey: Prentice-Hall, Inc., 1968.

Brandon, D. H. *Management Standards for Data Processing.* Princeton: D. Van Nostrand Company, Inc., 1963.

Brown, Harry L. *EDP for Auditors.* New York: John Wiley & Sons, Inc., 1968.

Buchholz, Werner. *Planning A Computer System.* New York: McGraw-Hill Book Company, 1962.

Canning, Richard G. and Sisson, Roger L. *The Management of Data Processing.* New York: John Wiley & Sons, 1967.

Chapin, Ned. *An Introduction to Automatic Computers.* Princeton, New Jersey: D. Van Nostrand Company, Inc., 1963.

Davis, Gordon B. *Computer Data Processing.* New York: McGraw-Hill Book Company, 1969.

----. *Auditing and EDP.* New York: American Institute of Certified Public Accountants, 1968.

Dearden, John and McFarlan, F. Warren. *Management Information Systems: Text and Cases.* Homewood, Illinois: Richard D. Irwin, Inc., 1966.

Desmonde, William H. *Real-time Data Processing Systems.* Englewood Cliffs, New Jersey: Prentice-Hall, Inc., 1964.

Elliott, C. Orville and Wasley, Robert S. *Business Information Processing Systems.* Homewood, Illinois: Richard D. Irwin, Inc., 1968.

Fisher, F. P. and Swindle, G. F. *Computer Programming Systems.* New York: Holt, Rinehart, and Winston, 1967.

Gentle, Edgar C., Jr., ed. *Data Communications in Business.* New York: American Telephone and Telegraph Company, 1966.

Gruenberger, Fred, ed. *Critical Factors in Data Management.* Englewood Cliffs, New Jersey: Prentice-Hall, Inc., 1969.

Head, Robert V. *Real-time Business Systems.* New York: Holt, Rinehart and Winston, Inc., 1964.

Hodge, Bartow and Hodgson, Robert N. *Management and the Computer in Information and Control Systems.* New York: McGraw-Hill Book Company, 1968.

International Business Machines, Inc. *Data Communications Primer.* Form C20-1668.

----. *Form and Card Design.* Reference Manual C20-8078.

Johnson, Richard A., Kast, Fremont T., and Rosenzweig, James E. *The Theory and Management of Systems.* New York: McGraw-Hill Book Company, 1967.

Kanter, Jerome. *The Computer and the Executive.* Englewood Cliffs, New Jersey: Prentice-Hall, Inc., 1967.

Laden, H. N. and Guildersleeve, T. R. *System Design for Computer Applications.* New York: Wiley, 1967.

Lecht, Charles Philip. *The Management of Computer Program-*

ming Projects. New York: American Management Association, Inc., 1967.

Martin, James. *Design of Real-time Systems.* Englewood Cliffs, New Jersey: Prentice-Hall, Inc., 1965.

Martin, Wainright E., Jr. *Electronic Data Processing: An Introduction.* Rev. ed. Homewood, Illinois: Richard D. Irwin, Inc., 1965.

McMillan, Claude, Jr. and Gonzalez, Richard F. *Systems Analysis: A Computer Approach to Decision Models.* Rev. ed. Homewood, Illinois: Richard D. Irwin, Inc., 1968.

Orlicky, Joseph. *The Successful Computer System: Its Planning, Development, and Management in a Business Enterprise.* New York: McGraw-Hill Book Company, 1968.

Porter, W. Thomas. *Auditing Electronic Systems.* Belmont, California: Wadsworth Publishing Company, 1966.

Prince, Thomas R. *Information Systems for Management Planning and Control.* Homewood, Illinois: Richard D. Irwin, Inc., 1966.

Rosen, Saul, ed. *Programming Systems and Language.* New York: McGraw-Hill Book Company, 1967.

Sanders, Donald H. *Computers in Business: An Introduction.* New York: McGraw-Hill Book Company, 1968.

----. *Computers and Management.* New York: McGraw-Hill Book Company, 1969.

Schoderbek, Peter B., ed. *Management Systems.* New York: John Wiley & Sons, Inc., 1967.

"Sorting Symposium," *Communications of the A.C.M.,* 6: No. 5 (May 1963).

Sprague, Richard E. *Electronic Business Systems: Management Use of On-line—Real-time Computers.* New York: The Ronald Press Company, 1962.

Tonley, Helen. "Information Coding Techniques." *Scientific Business,* Spring 1963, p. 39.

Valle and Reece. "Guidance Control for Business Systems." *International Business Automation,* December 1965, p. 26.

Wegner, Peter. *Programming Languages, Information Structures, and Machine Organization.* New York: McGraw-Hill Book Company, 1968.

Wofsey, Marvin M. *Management of Automatic Data Processing.* Washington, D. C.: Thompson Book Company, 1968.

The text of this book was set by
the Fototronic process in nine-point
Optima, a sans serif face designed
in 1958 by Hermann Zapf. Heads are
variations of Bailey Grotesk.

Composition was done by Spartan
Typographers of Oakland, California.

This book was printed on Mead
Meadium vellum offset paper by the
Kingsport Press of Kingsport, Tennessee.